# Contents

D1795252

Insight and Perspective Ltd, 701 Stonehouse Park, Sperry Way, Stonehouse, Glos, GL10 3UT

www.insightandperspective.co.uk

First published 2019

10 9 8 7 6 5 4 3 2 1

ISBN 13: 978-1-912190-05-8

Designed and typeset by Wooden Ark
Printed by TJ International, Padstow, Cornwall, UK

**Acknowledgements**
The author and publishers would like to thank Sam Rudd, Christine Wise, and Andy Leeder for their contributions when developing the Fieldwork in Action series.
The author and publishers would like to thank the following for permission to use the photographs/copyright material:
Front cover and p112 Field Studies Council; p16 OS Mapping of Shrewsbury © Crown copyright 2018 OS 93823011: p72 Josie Owen;
p39 and pp92-93 parallel.co.uk; p39 www.police.uk;
p57 and 115 hydrograph www.riverlevels.uk/esk.
All other photographs Andy Owen.

The publishers have made every effort to trace the copyright holders. If they have inadvertently overlooked any they will be pleased to acknowledge these at the first available opportunity.

The teaching content of this resource is endorsed by OCR for use with specification OCR GCSE (9-1) Geography A Geographical Themes (J383) and specification OCR GCSE Geography (9-1) Geography B for Enquiring Minds (J384). In order to gain OCR endorsement, this resource has been reviewed against OCR's endorsement criteria. This resource was designed using the most up to date information from the specification. Specifications are updated over time which means there may be contradictions between the resource and the specification, therefore please use the information on the latest specification and Sample Assessment Materials at all times when ensuring students are fully prepared for their assessments.
Any references to assessment and/or assessment preparation are the publisher's interpretation of the specification requirements and are not endorsed by OCR. OCR recommends that teachers consider using a range of teaching and learning resources in preparing learners for assessment, based on their own professional judgement for their students' needs. OCR has not paid for the production of this resource, nor does OCR receive any royalties from its sale. For more information about the endorsement process, please visit the OCR website, www.ocr.org.uk.

# Why do geographers do fieldwork?

Fieldwork is a really important part of geography. Designing a fieldwork investigation encourages you to be inquisitive and pose questions. Each fieldtrip teaches you useful skills about data collection and data analysis. After a fieldtrip you learn about different styles of maps and graphs – and why some are more suitable than others. Fieldwork also helps you understand the importance of evidence in constructing a persuasive piece of writing. Perhaps the most important skill that you will learn is how to evaluate your fieldwork. What were the strengths and limitations of your data collection or the ways you presented the data?

As a GCSE student you must do **two** fieldwork enquiries.
- One will investigate an aspect of physical geography such as rivers or coasts.
- The other will investigate an aspect of human geography such as urban or economic geography.

# How to use this book

This book is designed to support you throughout your fieldwork enquiries and in preparing for the examination which assesses fieldwork skills and enquiries.

There are four parts to the book:
- Part 01 The enquiry process;
- Part 02 Fieldwork in human environments;
- Part 03 Fieldwork in physical environments;
- Part 04 Preparing for your fieldwork assessment.

## Part 01 The enquiry process

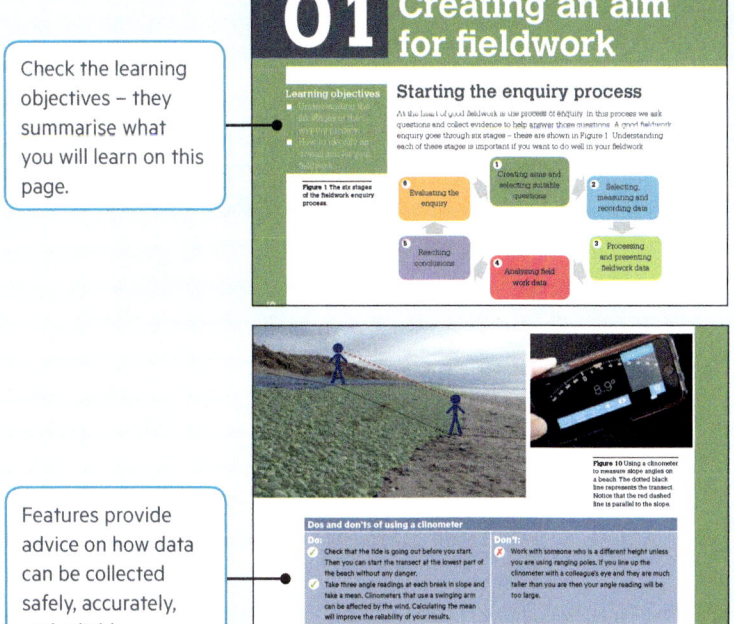

Check the learning objectives – they summarise what you will learn on this page.

Part 01 has 6 chapters – one for each stage of the enquiry process. Use these chapters as you plan your fieldwork and also when you get back from the fieldtrip to help you process the data, draw your graphs, and reach your conclusions.

Each part of the book has a different colour so you can find your way around easily.

Features provide advice on how data can be collected safely, accurately, and reliably.

# Part 02 Fieldwork in human environments

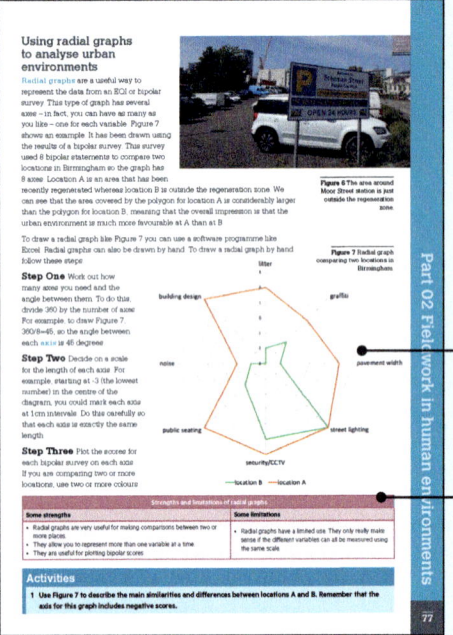

### Using radial graphs to analyse urban environments

Radial graphs are a useful way to represent the data from an EQI or bipolar survey. This type of graph has several axes – in fact, you can have as many as you like – one for each variable. Figure 7 shows an example. It has been drawn using the results of a bipolar survey. This survey used 8 bipolar statements to compare two locations in Birmingham so the graph has 8 axes. Location A is an area that has been recently regenerated whereas location B is outside the regeneration zone. We can see that the area covered by the polygon for location A is considerably larger than the polygon for location B, meaning that the overall impression is that the urban environment is much more favourable at A than at B.

To draw a radial graph like Figure 7 you can use a software programme like Excel. Radial graphs can also be drawn by hand. To draw a radial graph by hand follow these steps.

**Step One** Work out how many axes you need and the angle between them. To do this, divide 360 by the number of axes. For example, to draw Figure 7, 360/8=45, so the angle between each axis is 45 degrees.

**Step Two** Decide on a scale for the length of each axis. For example, starting at -3 (the lowest number) in the centre of the diagram, you could mark each axis at 1cm intervals. Do this carefully so that each axis is exactly the same length.

**Step Three** Plot the scores for each bipolar survey on each axis. If you are comparing two or more locations, use two or more colours.

**Figure 6** The area around Moor Street station is just outside the regeneration zone.

**Figure 7** Radial graph comparing two locations in Birmingham

litter
graffiti
pavement width
street lighting
security/CCTV
public seating
noise
building design

— location B — location A

**Strengths and limitations of radial graphs**

| Some strengths | Some limitations |
|---|---|
| • Radial graphs are very useful for making comparisons between two or more places. | • Radial graphs have a limited use. They only really make sense if the different variables can all be measured using the same scale. |
| • They allow you to represent more than one variable at a time. | |
| • They are useful for plotting bipolar scores. | |

### Activities

1 Use Figure 7 to describe the main similarities and differences between locations A and B. Remember that the axis for this graph includes negative scores.

77

Part 02 is about fieldwork investigations in a human environment.

Step-by-step guides help you collect data.

Maps or graphs help you practise analysing patterns or evaluating techniques.

Strengths and limitations help you evaluate fieldwork techniques. Learning how to evaluate your fieldwork is a really important skill – testing whether or not you can evaluate your fieldwork is one of the main points of the fieldwork exam questions.

Activities – so you can check you have understood what you have learned.

# Part 03 Fieldwork in physical environments

Part 03 is about fieldwork investigations in a physical environment.

These bullet points give ideas about collecting data in your own fieldwork. They will also help you prepare for the exam and answer questions about fieldwork in an unfamiliar environment. This is because your fieldwork may have been in a coastal location but you may still be expected to answer questions about fieldwork on a river.

Key fieldwork terms are in a bold coloured font – you can look them up in the glossary.

Photos help you to imagine how data can be collected.

## Chapter 10
# Investigating river environments

### Learning objectives
■ How to investigate processes in a river channel

### Investigating river processes

The processes of erosion, transportation, and deposition all happen in river channels like the one shown in Figure 1. Sediment is transported down the river. Where water has less energy, as in the shallow water at X, sediment is deposited. You can see evidence of this in the photograph – pebbles have been deposited in a river beach, also known as a point bar or slip-off slope. Erosion also occurs in meandering river channels. Erosion tends to occur on the outside bend of the meander where water flows with greater energy - forming a river cliff like the one at Y in Figure 1.

**Figure 1** A meandering river channel.

We could investigate the processes in this river channel by:
■ measuring the depth and width of the river channel and drawing a cross section.
■ measuring the velocity of the water and calculating discharge (see pages 84-85).
■ analysing the size of pebbles on the river's bed and point bar to find evidence of sorting.
■ taking photos or making field sketches and annotating the main features (see pages 32-34).

### Creating a cross section of a river channel

To create a cross section of the river channel you will need to follow these steps.

**Step One** Measure the width of the river channel. Divide this by 10 to create 10 evenly spaced sample points. In Figure 2 the channel is 3m across, so the samples are 30cm apart.

**Step Two** Stretch a line tightly across the river. Make sure it is level.

**Step Three** Measure downwards, from the line to the ground, at each sample point. In Figure 2, measurements were made every 30cm on each side of the river as well as into the river channel. Record the distance from the line to the surface of the water.

**Figure 2** How to take data readings to create a cross section across a river channel. Shropshire Hills AONB.

108

# Part 04 Preparing for your fieldwork assessment

### Reflecting on your enquiries

The most useful way of preparing yourself for the fieldwork assessment is to think critically about what you did and why you did it during each of your enquiries. Remember, there are no marks for describing your fieldwork. Instead, think about each step of the **enquiry process** and **evaluate** what went well and what wasn't so good.

There are some techniques you can use to help with your evaluation. For example, you could use a fish bone diagram (page 124) or a Lotus Diagram (pages 122-123) to help you evaluate a particular part of your investigation. To get an overview of your whole fieldwork experience, you could make a large copy of Figure 4. Use the questions in Figure 4 to think about specific things that you did during your fieldwork that worked well or not so well. Fill as much of the table as you can with specific details of these strengths and weaknesses.

**Learning objectives**
- How to make revision active – doing things to help you learn rather than just reading your notes.

**Figure 4** You should reflect on each stage of your fieldwork enquiries

| Stage of the enquiry process | Questions to ask yourself | Strengths of my fieldwork | Weaknesses of my fieldwork |
|---|---|---|---|
| 1 Creating aims and selecting suitable questions | Did I have SMART aims? (See page 9.) Where did I decide to collect my data? Looking back, were these the best places? | | |
| 2 Selecting, measuring and recording data | Did I use systematic, random, stratified, or opportunistic sampling? Can I justify my choice? What might have happened if I had used a different sampling strategy? | | |
| 3 Processing and presenting fieldwork data | Did I collect discrete or continuous data? Did I choose the most suitable methods to represent this data? | | |
| 4 Analysing fieldwork data | Was I able to see patterns, trends, and correlations in my data? Would trends have been clearer if I had used more sample points? Was I able to use the data to draw a map? If not, why not? | | |
| 5 Reaching conclusions | Was I able to reach firm conclusions from my data? If not, why not? Could I rely on any secondary data? | | |
| 6 Evaluating the enquiry | Did everyone use the data collection method in the same way so that I got reliable results? What might have happened if I had collected data at a different time of day or week? | | |

121

---

Part 04 describes how fieldwork is examined. It includes ideas about how you can review and evaluate your own fieldwork enquiries.

# Glossary

### Glossary

**Accuracy** How close a measurement is to the true value of the object being measured.
**Aim** What you hope to achieve or prove through fieldwork.
**Altitude** Height above sea level.
**Analysis** Process in which you make sense of the evidence.
**Anemometer** Fieldwork equipment used to measure wind speed.
**Annotation** Text added to a photo or artwork to describe and highlight key features.
**Anomaly** Data that does not fit into the general pattern or trend.
**Assess** To use data and evidence to make a judgement.
**Average** The central point of a data set. The average can be expressed in three ways.
**Mean** Calculated by adding up all data in a set of data to find the total then dividing that total by the number of items of data.
**Median** The middle value when all data in a set of data is arranged in rank order.
**Mode** The most frequently occurring value in a set of data.
**Axis/y-x** The x-axis is the horizontal base line of a graph. The y-axis is the vertical line.
**Bar chart** A way to represent data. Values are shown using vertical columns or horizontal bars.
**Divided bar** A rectangle which has been divided into segments to represent data in the form of percentages.
**Base map** A simple, outline map of a place. Data recorded during fieldwork can be added to the base map.
**Beach profile** A cross section showing the landscape of a beach, both above the water and below it.
**Belt transect (continuous & interrupted)** A sampling strategy. Data is collected along a line (or transect) usually by using a quadrat.
**Bias** A tendency to be positive or negative about something or someone.
**Big data** Large sets of data and databases of secondary data such as the National Census.
**Bipolar survey** Data collection method in which people express an opinion by choosing from opposite pairs of statements.
**Break in slope** A place along a transect where the gradient of the slope suddenly changes.
**Causality** The relationship between something that happens and the thing that makes it happen.
**Census** A national database about the UK population that is collected every 10 years.
**Choropleth** Type of map where darker shading is used to represent higher data values.
**Clinometer** Fieldwork equipment used to measure gradient / slope angle.

**Closed questions** Questions with specific set answers so the person answering has to choose the answer they must agree with.
**Clustered** Group of similar things gathered or occurring closely together.
**Conclusion** A summary statement or judgement that summarises the findings of a fieldwork enquiry.
**Conflict matrix** A way to record potential conflicts, for example, between different uses of a place or space.
**Connection** The link that exists between one set of data and another.
**Control group** A group of data that acts as a benchmark. A researcher can compare the results from the sampled group to the control group to help identify the impact of different variables.
**Control readings** Measurements that are taken so that the effects of one variable can be isolated from an investigation.
**Correlation** The connection that links two sets of data (or variables) together.
**Negative correlation** Values in one set of data increase as values in the other set of data decrease.
**Positive correlation** Values in both sets of data increase at the same time.
**Counting** To record how many whole units there are.
**Cross section** Represents the shape or profile of a feature using measurements of the distances and depths from a horizontal line.
**Data** Facts and statistics collected together, for example, as evidence and for analysis.
**Big data** A very large and complex set of data.
**Bivariate data** Two sets of data that are connected in some way.
**Census data** Data about the whole population which is collected every 10 years.
**Continuous data** Values that can be measured and recorded to one or more decimal point.
**Discrete data** Values that can be counted and recorded as whole numbers.
**Primary data** Data that is collected first hand.
**Qualitative data** Evidence that is collected as words, opinions, or images.
**Quantitative data** Evidence that is collected as number values.
**Secondary data** Data that has already been published in another source.
**Dataset** A group of values (set of data) that is often presented in a table.
**Desire line map** Uses thin straight lines to show how places are linked together.
**Discharge** A measure of water flowing in a river. Discharge is calculated by multiplying cross sectional area of the channel by the river's velocity.

**Dispersion graph** A way to represent data. Data is plotted to show the range of values.
**Dot map** Uses individual dots to show the exact location of similar features on a map.
**Elaborate** To write an extended answer that links ideas and often uses connectives like 'so', 'therefore', 'because'.
**Enquiry process** The process by which a student collects and analyses evidence in order to make a decision or prove/improve an aim.
**Enquiry question** A question that can be posed at the beginning of a fieldwork investigation to give the investigation an aim.
**EQI (Environmental Quality Index)** A technique which uses detailed criteria (statements) to assess the quality of our surroundings.
**Evaluate** To weigh up the strengths (or advantages) against the weaknesses (or disadvantages).
**Evaluation** A process in which you weigh up strengths and limitations. Evaluation is an important stage in your fieldwork enquiry in which you assess how well your fieldwork has gone. For example, you can evaluate the methods you used to collect or represent data.
**Evidence** Information that can be used to investigate an aim, test a hypothesis, or support an argument. Evidence may include quantitative or qualitative data.
**Extrapolation** A process by which a graph is used to find a data value that lies outside (or beyond) other data values.
**Field sketch** A sketch made during the fieldwork which records evidence about the most important features of a fieldwork location.
**Fieldwork** Practical work undertaken in natural and urban environments to investigate geographical questions and hypotheses.
**Flow line map** A way to represent data. The amount of a flow (such as traffic or pedestrians) is shown using a proportional arrow.
**Flow meter** A piece of equipment to measure the velocity of the flow of a river.
**Footfall** A measurement of the number of pedestrians.
**Frequency** The rate at which something happens over a particular period of time.
**Geological map** A way to represent data. Rock types are shown on a map by using different colours.
**GIS (Geographic Information System)** A system which captures, stores, analyses, and presents types of geographical data.
**Google street view** A feature in Google Maps and Google Earth that provides panoramic views from positions along many streets in the world.

Glossary

126

---

Use the Glossary to check you are using terms correctly when you are writing about fieldwork.

# PART 01

# Chapter 1

# Creating an aim for fieldwork

## Learning objectives

- Understanding the six stages of the enquiry process.
- How to identify an overall aim for your fieldwork.

**Figure 1** The six stages of the fieldwork enquiry process.

# Starting the enquiry process

At the heart of good fieldwork is the process of enquiry. In this process we ask questions and collect evidence to help answer those questions. A good fieldwork enquiry goes through six stages – these are shown in Figure 1. Understanding each of these stages is important if you want to do well in your fieldwork.

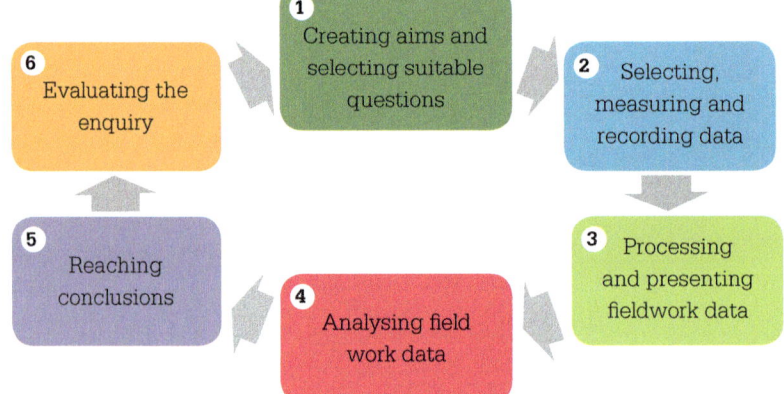

## Choosing an aim

You can't just start your fieldwork by going outside and measuring things or counting stuff. Fieldwork involves a lot of counting and measuring but this data collection has to have a purpose so you will need to give your fieldwork an **aim**. The aim sums up what you hope to achieve or prove through your fieldwork. There are three main ways to decide the aim of your fieldwork enquiry.

1 **Investigate a geographical process** to see how it works in the real world. For example, you could investigate the process of longshore drift on a beach, or the deposition of pebbles in a river. You could investigate the process of regeneration in a town or city.

2 **Survey people's opinions about a geographical issue**. This may be some controversial change that is affecting your local area such as the building of a new housing estate on a greenfield site.

3 **Investigate a geographical concept**. Concepts are the big ideas that help us make sense of the world around us. Concepts help us break down complex patterns and make connections. Figure 2 lists some useful concepts.

You also need to know where your fieldwork is going to take place. Will you be working in a river environment, a beach, or in a town? A little local knowledge may help you to identify an interesting issue that could be investigated. Thinking about the location and a process, issue, or concept, will help you create an aim for your fieldwork.

| Concept | What is this concept about? |
|---|---|
| Cause and effect | Geographical features and processes have positive and negative effects on the environment surrounding them. |
| Cycles and flows | Environments contain flows, for example, flows of water through a drainage basin, or flows of people through a city. We can investigate factors that influence the size and rate of these flows. |
| Inequality | Services such as sports facilities or social care are rarely distributed evenly. We can investigate whether one place, or one group of people, has more than its fair share of something. |
| Mitigating risk | All environments contain risks such as river or coastal floods. We can investigate how these risks can be reduced (or mitigated) or managed. |
| Place | Each place has features that give that place a unique character. This character gives the people who live in that place a sense of identity. |
| Sustainability | Sustainability is a measure of how well a place or environment is coping with social, economic, or environmental change. |

For example in Figure 3, we could investigate:

- a process like gentrification;
- an issue such as traffic congestion and pollution;
- the concept of risk.

**Figure 2** Geographical concepts that can be investigated through fieldwork.

Only after you have created an aim for the fieldwork can you give your investigation a title. A title could be a simple statement that reflects the aim of your investigation:

*An investigation of the issue of traffic congestion in Cowbridge Road.*

Alternatively, you could choose a title that that reflects the main question that you want to answer, such as:

*How can Cowbridge Road be made safer for pedestrians and cyclists?*

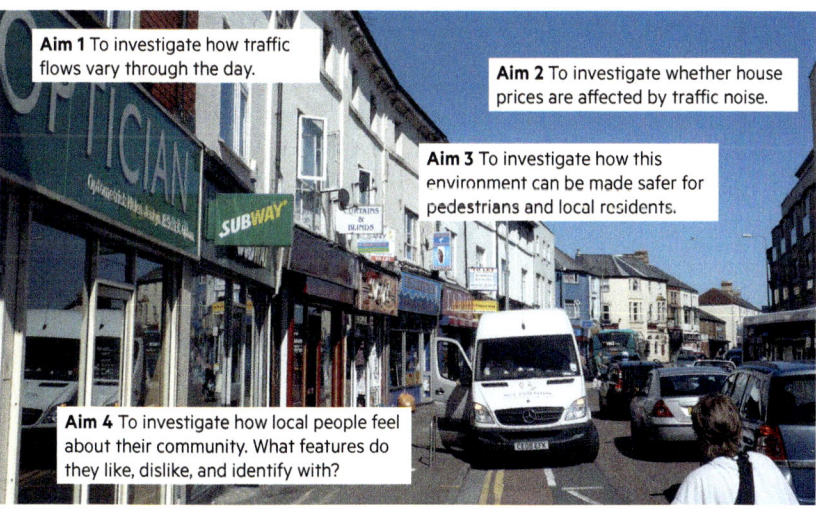

**Aim 1** To investigate how traffic flows vary through the day.

**Aim 2** To investigate whether house prices are affected by traffic noise.

**Aim 3** To investigate how this environment can be made safer for pedestrians and local residents.

**Aim 4** To investigate how local people feel about their community. What features do they like, dislike, and identify with?

**Figure 3** Which aim would you investigate here? Cowbridge Road East, Cardiff.

## Activities

1 **Study Figure 3. If you were able to visit this location on a fieldtrip:**
   a) how might counting the traffic help you answer one or more of these aims?
   b) what other data could you collect to help you answer each of these aims?
2 **Each of the aims in Figure 3 is connected to one of the concepts listed in Figure 2.**
   a) Match the aims to the concepts.
   b) Suggest one other concept that could be investigated here. Explain your choice.

## Learning objectives
- Creating a SMART aim.
- Asking enquiry questions.
- Setting a hypothesis.

**Figure 4** Dovedale in the Peak District National Park. Dovedale is a honeypot site that attracts thousands of visitors, especially at weekends. This is known as a honeypot site.

# What makes a good enquiry question?

You can use questions to break down your aim into manageable chunks but what makes a good enquiry question? To help you ask your own enquiry questions you could use the question starters that have been added to Figure 4.

Figure 5 gives three different styles of question.
- Questions on the left are useful but not very demanding because the answers are probably very straightforward.
- The questions in the middle are better. They are more useful and interesting as enquiry questions because they are about the relationship between two variables.
- The questions on the right are more interesting again and could provide excellent enquiry questions because you may find a variety of potential answers. However, this type of question needs to be phrased very carefully and you may need to gather a great deal of evidence to answer them fully.

**Figure 5** Styles of enquiry question.

| | increasing complexity of question ⟶ | | |
|---|---|---|---|
| | Questions that have a straightforward answer | Questions about connections | Questions that have a complex answer |
| **Examples of question starter** | When? Where? How many? Which? | How does? What effect would? | Who benefits? What ought? Who should? |
| **Examples of questions** | Which is the busiest shopping street? When is the busiest time of day? How many cars pass in 5 minutes? | How does wind speed vary with altitude? What effect would traffic noise have on house prices? | Who would benefit most if flood protection was improved? What ought to be done to make this community more sustainable? |

## Choosing a hypothesis

Just like a science experiment, you can sometimes use geography fieldwork to prove, or disprove, a hypothesis. A **hypothesis** is a provisional idea which can be proven to be correct, or incorrect, based on the factual evidence collected in your fieldwork. An example of a hypothesis that could be investigated in Figure 4 is:

*Footpath erosion is greater closest to the car park.*

Once you have chosen a suitable hypothesis for your fieldwork you should state the **null hypothesis**. This is what you would discover if the hypothesis was incorrect. For Figure 4, the null hypothesis would be:

*There is no difference in the amount of footpath erosion with increasing distance from the car park.*

## Setting a SMART aim

Is your aim a good one? Is it suitable? An aim shouldn't be too simple. For example, *'How many cars travel down this road in 5 minutes?'* is not an aim. This simple question is part of the broader aim of investigating patterns of commuter movement. However, an aim shouldn't be too complex either – otherwise you may not be able to achieve it. For example, *'Cowbridge Road East is a perfect example of a modern sustainable residential community'* would be very difficult to prove, or disprove, because the word 'perfect' is so difficult to measure. Your aim should be SMART – see Figure 6.

**Is your aim SMART?**
- **Specific** – does the aim have a clear focus?
- **Measurable** – can you collect and count data?
- **Achievable** – will it be safe? Will you need a team of people? Is data available?
- **Realistic** – have you got the right skills and/or equipment?
- **Timely** – will you have enough time during the fieldtrip to collect all of the data?

**Figure 6** Making your aim SMART.

## Take a virtual visit

Another good way to prepare for your fieldwork is to make a virtual visit. Use the internet to see the environment you will be working in before the day of the fieldtrip. This has a number of advantages. It will help you to:
- pose enquiry questions or set a hypothesis that relates to the actual place that you will visit;
- plan where you could collect the data;
- consider potential risks of the site and plan your visit so you stay safe.

### Activities

1 **Study Figure 4.**
   a) Write 10 enquiry questions using as many different question starters as you can.
   b) Evaluate your questions using Figure 5. Which are your best three questions? Are they SMART? Are they complex?

**Figure 7** Busy roads can be a risk. Only cross where it is safe.

# Risk assessment

You need to stay safe during your fieldwork, so it is important to think about any potential risks that you and your classmates might face. If you can identify the potential risks, then you can think about specific ways that the danger can be reduced or removed completely. For example, if you are working next to a busy road you must stay on the pavement. If you have to cross the road, then only do so at a safe location – preferably at a crossing point that is controlled by lights as in Figure 7. Figure 8 lists some potential risks you could face during fieldwork and ways these risks can be reduced.

| Potential risk | How this risk can be reduced |
|---|---|
| Being trapped on a beach by the rising tide. Risk of drowning. | Check the tide table before the visit. Do not work on the beach as the tide is rising. |
| Strong winds blow sand from the beach. Risk of eye injury. | Check the weather forecast before the visit. Do not work in sand dunes when there are strong onshore winds. |
| Rocks falling from cliffs above. | Do not work at the top of a beach immediately below a cliff. |
| Falling on a steep slope and suffering a serious injury to limbs or head. | Do not work close to the top of a cliff. Keep to paths on slopes. |
| Slipping on wet grass and twisting an ankle. | Wear walking boots rather than trainers. |
| Falling into a river and suffering an injury or hypothermia. Risk of drowning. | Do not run next to the river. Wear walking boots. Students working in a river must not go in above knees and always work in teams. Check the weather forecast before the visit. |
| Being hit by a car/cyclist when crossing the road. | Only cross roads where visibility in both directions is good or where there is a pelican or zebra crossing. |
| Trip hazards when walking through undergrowth. | Wear walking boots. Move slowly and carefully. |
| Low branches in a woodland. Risk of eye injury. | Keep to paths. Do not run. |
| Tunnelling in sand dunes. Risk of collapse and suffocation. | Do not play in sand dunes and never dig or tunnel. |

**Figure 8** Potential risks and how to reduce them.

## Think of others

As well as identifying potential risks, there are some things we can all do during a fieldtrip to keep safe. These tips should make your experience of fieldwork pleasant as well as safe.

- Listen to your teacher's instructions carefully.
- Work in pairs or small teams. Look after each other.
- Wear sensible clothing and shoes so you stay warm and safe.
- Be aware of hazards around you. Look out for warning signs, like Figure 9, if you are working in an environment that is potentially dangerous.

**Activities**

1 **Study Figure 10.**
   a) Identify three potential risks of doing fieldwork in this location.
   b) Suggest how each risk could be reduced.

**Figure 9** A sign warning of risk at the coast, West Wales.

**Figure 10** A beach at Happisburgh, Norfolk.

# Chapter 2

# How is evidence collected?

## Learning objectives

- Understanding the difference between qualitative and quantitative data.
- Understanding the difference between primary and secondary data.

## Types of data

In a fieldwork enquiry we need to collect data that will provide the evidence we need to answer our aims. We describe data as either **quantitative** or **qualitative**.

**Quantitative data** is something you can count, or measure. Some evidence that you can measure will need specialist equipment. For example you measure:

- wind speed with an **anemometer**;
- river velocity with a **flow meter**;
- slope angles with a **clinometer**.

However, lots of quantitative data can be simply counted such as the number of vehicles on a road, the number (and type) of shops, and the number of pedestrians in a shopping centre.

**Qualitative data** is something that helps us to describe a place or geographical issue with words or pictures rather than by measuring with numbers. It can include interviews, making field sketches, or taking photographs, videos, or audio recordings.

**Figure 1** Examples of quantitative and qualitative data.

| Quantitative data | Qualitative data |
|---|---|
| Depth and width of a river<br>Wind speed<br>House prices<br>The size of pebbles on a beach | Whether pedestrians feel safe in the city centre<br>Whether people prefer shopping in the high street or out of town<br>What different groups of people think about the risk of flooding |

## Activities

1   **Make a copy of Figure 1. Decide whether each statement below is quantitative or qualitative and then add it to the correct column.**

The number of wave crests that break each minute on a beach.

Photos of features of an urban environment that provide evidence of sustainability.

The reasons people move to live somewhere new.

A field sketch of a coastal landform.

The angle of slope of a beach.

Photos of features of a shopping centre that are liked/disliked by shoppers.

Noise (in decibels) of traffic from a road.

An audio recording of a busy shopping street.

Infiltration rates of water soaking into different soils.

Wind speeds at different locations up a hill.

## What is the difference between discrete and continuous data?

**Discrete data and continuous data** are two different types of quantitative data. **Discrete data** can be counted. We could count how many people are in a shopping street, or how many vehicles pass along a street. That would be discrete data. By contrast, **continuous data** is measured. We can measure how quickly the river is flowing, or how much noise the traffic is making. That would be continuous data. Distance, velocity and time are all examples of continuous data that are often measured in geography fieldwork.

1 The number and location of security cameras.

2 The length of time that people spend looking at the information.

3 The price of flats in the city centre.

4 The number of clothing and shoe shops.

5 Noise levels (in decibels).

6 The number of shoppers.

**Figure 2** Examples of discrete and continuous data that could be collected in a retail environment, Lancaster.

## What is the difference between primary and secondary data?

The data you collect during the fieldtrip, such as the number of vehicles passing, or the answers to a questionnaire, is known as **primary data**. Sometimes you will be able to answer your aim by only collecting primary data. However, a lot of data has already been collected and is available in books and on websites. This is **secondary data**. There are lots of forms of secondary data that might be useful to help you answer your fieldwork enquiry. With careful research, you can find:

- photographs that show your fieldwork location, both now and in the past - you could annotate these to show how the location has changed;
- blogs – you can analyse people's viewpoints;
- maps - for example, the Environment Agency Flood Map would be useful if your fieldwork is located in a town that is at risk of river or coastal flooding;
- tables of data – which you can use to draw your own graphs.

| Strengths and limitations of primary and secondary data | | |
|---|---|---|
| | **Some strengths** | **Some limitations** |
| **Primary data** | You know exactly how the data was collected. | Your sampling strategy may not be representative (see pages 14-15). |
| **Secondary data** | It's much quicker to download secondary data than collect your own primary data. | You may not know how or when it was collected and the figures may be out of date. |

## Activities

2 **Study Figure 2.**
   a) Sort the quantitative data into discrete and continuous types.
   b) Suggest three examples of qualitative data that could be collected here.

# Why do I need a sampling strategy?

Imagine you want to see how the amount of traffic varies throughout the day in a small commuter town. It wouldn't be possible to count every single vehicle travelling along every road for 24 hours. It would take too much time. Also, you would need your friends to agree to help you because you can't watch every road at the same time. So, instead of trying to collect every bit of data, you will need to collect a **sample** of data. If collected properly, your sample will accurately represent the patterns or trends shown in all of the data.

## What sampling strategies are available?

**Figure 3** Four commonly used sampling strategies.

There are various ways you can select your sample. Figure 3 summarises four main strategies that you could use.

| Sampling strategy | How it works | Examples – you collect data... |
|---|---|---|
| Systematic | Your sample is chosen at regular intervals. Every $n^{th}$ item is sampled. | • at points that are equally spaced (for example, every 10 metres) along a line (transect)<br>• every hour<br>• by interviewing the 10th, 20th and 30th person who passes you |
| Random | Your data is sampled at random intervals. Every item has an equal chance of being sampled. | • at places that have been chosen by using tables of random numbers to generate grid references<br>• in streets chosen by putting all street names into a bag and pulling out a sample of names at random |
| Stratified | Your data falls into two or more categories. You make sure that your sample reflects the proportions that fall in each category. | • by sampling plants and wind speeds in a sand dune. You choose 50% of your sampling points on slopes that face the sea and 50% of your sample points on slopes that face away from the sea<br>• you are sampling how students at your school feel about a geographical issue. 45% of the students are in Years 7 to 9, 40% are in Years 10 and 11, 15% are in the sixth form. You make sure that your questionnaire is given out in these proportions. |
| Opportunistic (also known as convenience sampling) | Your data is scarce so you take the opportunity to collect data whenever or wherever you can. | • by asking a questionnaire of every tourist you can find<br>• you intended to systematically sample downstream along the course of a river. However, a lot of the sites you wanted to sample were on private land so you sampled wherever there was a public footpath next to the river. |

## Sampling along a line

**Figure 4** By collecting data at regular intervals across this river channel the students are sampling systematically along a line.

In many fieldwork enquiries a hypothesis will try to investigate how something changes over distance. For example:

> *rivers get wider as you go downstream;*
>
> *noise reduces as you get further away from a busy road;*
>
> *house prices get higher as you get closer to the park.*

Each of these hypotheses could be investigated by sampling data along a line. You can use any of the four main sampling strategies while you work along a line. Sampling along a line is often described as a **transect**.

## Sampling across an area

People say that if you can map it, it's geography! This is because geographers are interested in how features are distributed. The patterns we see on a map are known as **spatial** patterns. If you sample along a line you are sampling in one dimension. You will see how a pattern changes with distance but you will not be able to see a spatial pattern. To see a spatial pattern (and then draw a map) you will need to sample in two dimensions. You could do this by:

- sampling along several lines, for example, four lines radiating out from the town centre towards the suburbs;
- using a map of a town to identify sampling locations at each grid intersection. This would be a form of systematic sampling.

## Which is the best sampling strategy for me?

Each sampling method has advantages and disadvantages so each method is more suitable in some situations than others.

- **Random sampling** provides an unbiased sample. If the full size of the thing that is being sampled is quite small (like shops in a town, or students in a year group) this method should create a sample that represents the whole accurately – we say that it is **representative** of the whole.
- **Systematic sampling** is much quicker and simpler than random sampling. It's a better method to use if the thing that is being sampled is quite large (like a questionnaire of shoppers, or plants along a line through an ecosystem). Systematic sampling across an area should generate enough data to draw a map.
- When the thing that is being sampled is very large (for example, pebbles on a beach), you can combine random and systematic sampling strategies (see pages 18-19).
- **Opportunistic sampling** is really a method that you choose because it is not practical to choose another method. It may be quicker or easier than other methods but the results may not be representative.

http://www.textfixer.com/numbers/random-number-generator.php One of many online tools that will generate random numbers for you.

---

### Activities

1  **Think about the sampling strategy you used in your own fieldwork enquiry. Think about whether:**

- you were able to come to firm conclusions about changes over distance, or changes over time;
- it allowed you to spot spatial patterns;
- your data was representative of the population as a whole.

Now answer these questions:

a) What were the strengths and limitations of the sampling strategy that you used?

b) How could you have improved your own sampling?

# Ordnance Survey (OS) maps

You can use an **Ordnance Survey (OS) map** to design your sampling strategy before the fieldtrip. Figure 5 is an extract from a 1:50,000 map which means that 1cm on the map is equal to 50,000cm (500m) on the ground. A 1:50,000 scale map is useful for planning fieldwork in a town. For coastal or rivers fieldwork a 1:25,000 map is more useful because it shows more detail.

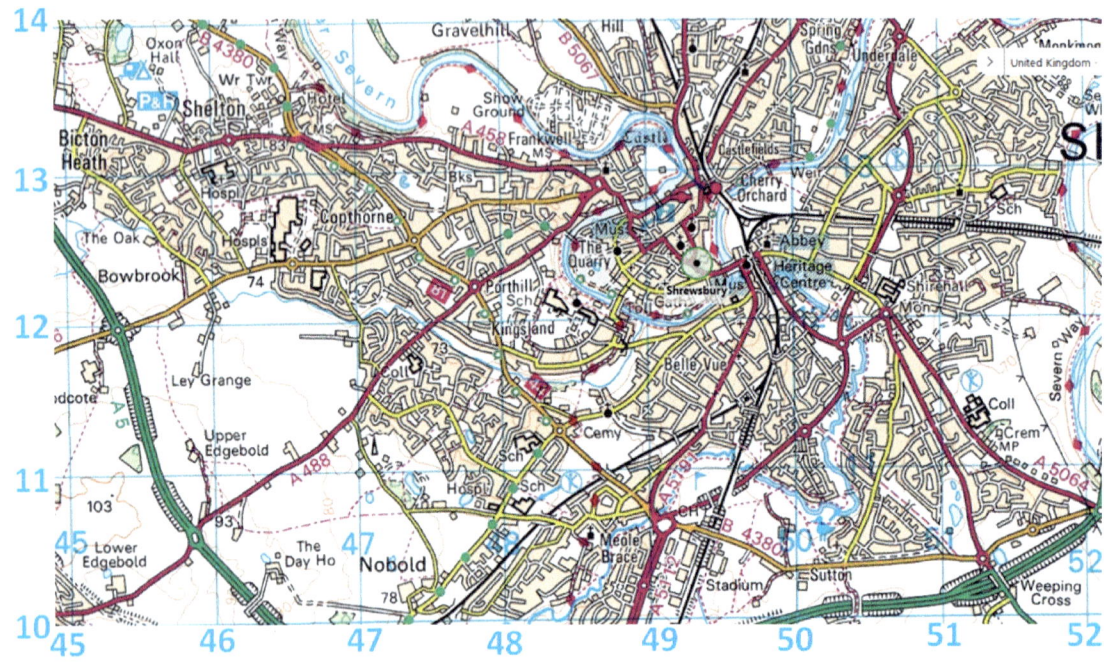

**Figure 5** 1:50,000 OS map extract of Shrewsbury.

## Using an OS map to sample along a line

You can use an OS map to plan the route of a **transect** (or cross section). For example, in Figure 5, we could sample traffic, noise, and house prices at regular intervals along the A458 to investigate how and why house prices vary with distance from the town centre. First, use the map to calculate the total length of the transect. It is about 4,500 metres from the edge of the town in grid square 4513 to the town centre, so we could sample at regular intervals of 500 metres to create a sample of 10 sites with one sample point at each end of the transect.

## Sampling across an area

The light blue grid lines on an OS map divide your fieldwork area into a grid. You can use this to design a systematic sampling strategy that will give two dimensional (2D) data that can be represented on a map later. For example, we could use Figure 5 to design a systematic sampling strategy for Shrewsbury using the following steps.

**Step One** Design an **EQI (Environmental Quality Index)** survey (see pages 28-29) that can be used to investigate how quality of life or access to services varies across the town.

**Step Two** Identify each grid square that contains part of Shrewsbury. There are 24 out of 28 grid squares in Figure 5. You could work in teams and allocate 4 squares to each team.

**Step Three** Carry out one EQI survey within each grid square.

## Checking access

OS maps are also useful for checking that you will have access to places that you would like to survey. In a river survey, for example, students often plan to sample at regular intervals along the stream. However, in reality, some sample points cannot be accessed because they are on private land. This means that survey points are chosen by opportunistic sampling because students actually sample the river where they have access. In Figure 5 you can see that the stream in grid square 5011 has a public footpath next to it so access would be possible.

## Grid references

Grid references are useful to identify sample sites, for example, the location of an EQI survey, a field sketch, or a photograph. You can give two types of grid reference.

- A four figure grid reference identifies a complete grid square.
- A six figure grid reference pinpoints a single place within one grid square.

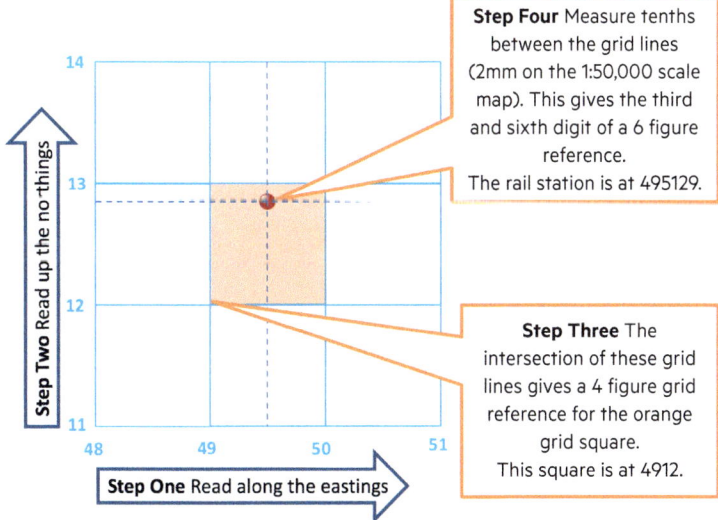

**Step Four** Measure tenths between the grid lines (2mm on the 1:50,000 scale map). This gives the third and sixth digit of a 6 figure reference.
The rail station is at 495129.

**Step Three** The intersection of these grid lines gives a 4 figure grid reference for the orange grid square.
This square is at 4912.

**Step Two** Read up the no things

**Step One** Read along the eastings

**Figure 6** How to give four and six figure grid references.

### Activities

1 **Design a transect with 10 regularly spaced sampled points that starts at Lower Edgebold (in the bottom left of the map) and ends at the train station.**
   a) Give a six figure reference for the start of the transect.
   b) How long is the transect?
   c) How far apart are the sample points?
   d) How many sample points are in the countryside before the edge of the town?

# Sampling data using a quadrat

A **quadrat** is a simple square frame that you place on the ground. It gives you a shape that you can sample inside. The frame is usually quite small – often 50cm by 50cm (which is a quarter of a square metre) - like the one shown in Figure 7. However, you can also use tent pegs and string to mark out a larger frame – say 2 metres by 2 metres. The frame is often divided into smaller squares with wire or string to make it easier to identify sample locations within the square.

Quadrats are used to help sample a very large population by breaking down a large space into smaller, more manageable spaces – concentrating on the space within the frame of each quadrat. They are often used to sample plants (for example, in a sand dune ecosystem) or pebbles (on a beach).

There are two main ways to use a quadrat.

- **One** Sample objects within the quadrat using the points where the internal wires/strings cross – in other words, at the intersections. Use them systematically, for example, by sampling pebbles at every intersection, as in Figure 7.
- **Two** When you are sampling types of vegetation, it is usual to estimate the area covered by each type of plant within the quadrat, as in Figure 8. Do this by counting the squares (the spaces between the wires/strings). The quadrat in Figure 8 is divided into 100 squares so it is easy to estimate the percentage distribution of each plant as each square represents 1%. This style of quadrat works best with low growing plants. The quadrat in Figure 7 has five rows and five columns of spaces between the strings. That makes 25 small squares inside the quadrat. The quadrat represents 100% of the sample area, so each small square represents 4% of the whole (because 100 divided by 25=4). This style of quadrat could be used to estimate percentage area covered by taller plants.

Quadrats are often used to sample vegetation along a belt transect.

**Figure 7** A quadrat being used to sample pebbles on a beach. The stone under each intersection was measured.

## Control groups

Quadrats are often used to measure how vegetation changes in relation to another variable, for example, how vegetation varies:

- across a sand dune with increasing distance from the sea;
- across a footpath with variable amounts of trampling.

When you are looking for variations like this it is a good idea to see what the vegetation is like by sampling in an area unaffected by the variable. We call this a **control group**. It allows you to compare the results of the area that you are investigating (for example, the plants that have been trampled on the footpath) with the normal vegetation of the area.

**Figure 8** Estimating areas within the frame of the quadrat. The percentage of the quadrat occupied by each species of plant was estimated. Three different species of plant can be identified in this quadrat. The second image has been tinted to show the percentage area covered by each plant more clearly.

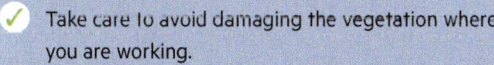

### Dos and don'ts of using quadrats

**Do:**

- ✓ Select locations for the quadrat using a systematic, random, or stratified sampling technique.
- ✓ Take care to avoid damaging the vegetation where you are working.

**Don't:**

- ✗ Throw the quadrat over your shoulder. This doesn't make the sampling random and it could be dangerous!
- ✗ Select pebbles from within the quadrat at random by just picking some out. Your eye will be drawn to the larger, or more interesting, looking pebbles. This means your sample won't be representative.

| Strengths and limitations of using quadrats | |
| --- | --- |
| Some strengths | Some limitations |
| Quadrats are useful for sampling a very large population (like pebbles or plants) that is spread over a large area. You can use quadrats to record the presence or absence of a variable (whether it is in the quadrat or not) and also the frequency of a variable (how much of the variable is present). Quadrats are relatively cheap pieces of equipment and simple to use. | It can be difficult to be accurate about the percentage of vegetation in a quadrat if the plants are different heights. Taller plants grow over shorter ones so it is possible to underestimate the percentage of the quadrat filled by smaller plants. |

# Using fieldwork equipment

A lot of data can be collected without the need for any special equipment. You will not need special equipment to collect qualitative data through **bipolar surveys**, **Likert surveys**, and **questionnaires**, although you will need to design a **data collection sheet** before the fieldwork begins (see pages 22-23).

Quantitative data that is continuous will need to be measured and some will need specialist fieldwork equipment. Wind speed, for example, needs to be measured using an **anemometer**, like the one in Figure 9. However, other measurements are easier to make. For example, in the river fieldwork seen in Figure 4 on page 14:

■ the width of the river is measured using a measuring tape;

■ the depth of water is measured using a metre rule;

■ the velocity of river flow can be calculated by floating a dog biscuit downstream for 10 metres and using the stop watch app on your smart phone.

**Figure 9** An anemometer is used to measure wind speed.

## Using a clinometer

A **clinometer** is used to measure angles from the horizontal so, in fieldwork, it can be used to measure the angle of a slope. A traditional clinometer includes a swinging arm or weight. As the clinometer is tipped up to the same angle as the slope, the weight points vertically downwards so that you can measure the angle. Alternatively, you could download a free clinometer app onto your smartphone, like the one shown in Figure 10.

To measure the slope of a beach, like the one in Figure 10, you will need:

■ a clinometer (or an app on your smartphone);

■ two people;

■ a long measuring tape;

■ two ranging poles (although these are not essential);

■ a data recording sheet.

**Step One** Set up a transect, up and down the beach, using the long measuring tape.

**Step Two** Identify the **breaks in slope** along the transect. These are places where the gradient of the slope suddenly changes. The photograph in Figure 10 has been coloured so that you can see the breaks in slope.

**Step Three** Start measuring at the bottom of the slope. If you do not have ranging poles, work with someone who is the same height as you. Get your partner to stand at the next break in slope. Hold the clinometer up to your eye and look along it into your partner's eyes. The angle recorded by the clinometer is the angle of the slope.

**Step Four** Record the angle (in degrees) and the distance between the start of the transect and the first break in slope (in metres) on your data recording sheet.

**Step Five** Later you can present your data as a line graph using a ruler and protractor. This graph will represent the shape of the **beach profile.**

**Step Six** Move up the transect to the first break in slope. Your partner should also move up the transect and stand at the next break in slope. Take the measurements and repeat this process until you have completed the transect.

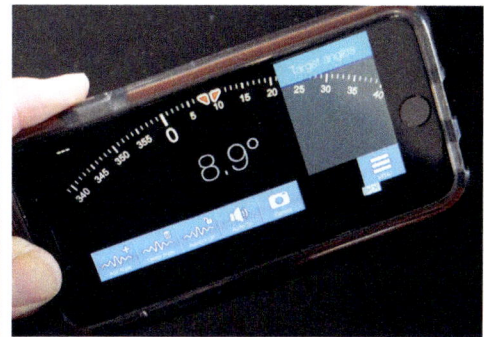

**Figure 10** Using a clinometer to measure slope angles on a beach. The dotted black line represents the transect. Notice that the red dashed line is parallel to the slope.

## Dos and don'ts of using a clinometer

**Do:**

✓ Check that the tide is going out before you start. Then you can start the transect at the lowest part of the beach without any danger.

✓ Take three angle readings at each break in slope and take a mean. Clinometers that use a swinging arm can be affected by the wind. Calculating the mean will improve the reliability of your results.

**Don't:**

✗ Work with someone who is a different height unless you are using ranging poles. If you line up the clinometer with a colleague's eye and they are much taller than you are then your angle reading will be too large.

| Site | Distance (metres) between sites | Angle (°) between sites |
|------|--------------------------------|-------------------------|
| 1 - 2 | 10 | 5 |
| 2 - 3 | 22 | 8 |
| 3 - 4 | 20 | 10 |
| 4 - 5 | 26 | 15 |
| 5 - 6 | 22 | 8 |
| 6 - 7 | 30 | 6 |
| 7 - 8 | 40 | 10 |
| 8 - 9 | 50 | 15 |
| 9 - 10 | 20 | 20 |
| 10 - 11 | 40 | 5 |
| 11 - 12 | 30 | 3 |

**Figure 11** This data was collected by students measuring slope angles on a beach in the UK.

## Activities

1 Present the information from Figure 11 as a line graph to show the beach profile. You will need graph paper and a protractor to do this accurately. Plot distance from the sea on the horizontal axis. For sites 1-2 draw a straight line of 10 metres length from the origin (0) at an angle of 5°. From the end of this line draw a line of 22 metres length at an angle of 8°. Continue until all the data is plotted.

2 Annotate your line graph to show significant features of the beach profile, such as breaks in slope.

Learning objectives
- Knowing when to use a pilot survey.
- Understanding how to design data collection sheets.

# Designing data collection sheets

Now you know what data you want to collect there is one more important job to do before you go on your fieldtrip. You will need a quick and methodical way to record the evidence that you collect on your fieldtrip – so you need to design **data collection sheets**.

Your data collection sheet needs to be easy to use. Always record when and where the data was collected at the top of the sheet. Data can be broken down into categories (for example, size of pebbles, or age of people) and you can then add boxes or tables to your sheet so you can record results in each category. Using a **tally mark**, like Figure 12, is a good way to collect information that you are counting.

Continuous data, such as wind speed and noise levels, can vary considerably as you are measuring them. It's a good idea, therefore, to create several boxes for these values on your data collection sheet, like in Figure 14. Take 5 readings, one every 30 seconds. Then take an average (mean).

Lots of students start a questionnaire, or bipolar survey, by recording the gender and age of the person they are talking to but this often isn't necessary. You should only do this if age or gender are factors that you want to analyse as part of your enquiry. For example, it would be essential to record the age of the respondents in your questionnaire if your enquiry question was 'Are older people more aware of risk in the environment than younger people?'.

**Traffic survey**

Location ..High Street......

Date ..19/05.. Time ..11.4m...

| Type | Tally | Total |
|---|---|---|
| Cycles | IIII | 4 |
| Motorbikes | II | 2 |
| Cars | HHt HHt HHt IIII | 19 |
| Vans | HHt HHt | 10 |
| Lorries | HHt I | 6 |
| Buses | HHt | 5 |

Figure 12 Use tally marks when you are counting things.

**Beach survey data collection sheet**

Date ......

Site 1

| Pebble sizes (mm) | Number of pebbles |
|---|---|
| Less than 10 | |
| 10-20 | |
| 25-40 | |
| 40-50 | |
| More than 50 | |

Figure 13 There are three errors in this data collection sheet. Can you spot them?

Location of noise readings ...............

Time they were taken ........................

| Readings (one taken every 30 seconds) | Noise level (dB) |
|---|---|
| 1 | |
| 2 | |
| 3 | |
| 4 | |
| 5 | |
| Total | |
| Average (total/5) | |

Figure 14 Design your data collection sheet so that several readings can be taken because noise readings may vary a lot.

| Dos and don'ts of data collection sheets | |
|---|---|
| **Do:** | **Don't:** |
| ✓ Create a space at the top of the data collection sheet to record the time, date, and location of your survey.<br>✓ Use boxes, tables, and tally marks to collect data methodically and quickly.<br>✓ Each column in your data collection table must have a heading. Make sure you also record the units of measurement in the heading, for example, Depth of water (cm). | ✗ Record the age and gender of the people in your questionnaire unless you are going to use this information in your analysis.<br>✗ Create categories that overlap each other or which miss out some individuals. |

## Pilot surveys

If you are going to use a questionnaire (see pages 26-27) you will want to use some closed and some open questions. **Open questions** allow the person you are interviewing to say whatever they want in response. The following are examples of open questions that you might use in a fieldwork enquiry about place and identity.

1. Which features do you like best about this town?
2. What makes this town different from others you know well?
3. What gives this place its unique identity?

Using open questions like these have advantages but it is time consuming to record all the possible responses that people might give. So, it's a good idea to use a **pilot survey** with a few people to see what kinds of answers you are likely to get when you carry out the full survey. After the pilot survey, you can look at the answers that people gave and put them into categories. So, in a pilot survey, the open question *'Which features do you like best about this town?'* was asked. Imagine that:

- 5 people said friendly people;
- 4 people talked about history or historic buildings;
- 3 people said that good schools were an important feature of the town;
- 3 people said they liked the multicultural atmosphere;
- 1 person talked about locally owned (independent) shops.

As a result of this pilot survey you can now design a data collection sheet for your full questionnaire that looks like Figure 15.

**Figure 15** Using a pilot survey to create option boxes for the data collection sheet.

1. Which features do you like best about this town?
Tick up to 3 boxes.

| | |
|---|---|
| The people | ☐ |
| The town's history | ☐ |
| Buildings/architecture | ☐ |
| Local schools | ☐ |
| Cultural activities | ☐ |
| Multicultural atmosphere | ☐ |
| Independent shops | ☐ |
| Others | ☐ |

If others, please describe the feature(s) …

## Activities

1. **Study Figure 13. Identify three errors in the design of this data collection sheet.**
2. **Explain why it would be important to use a pilot survey to create the data collection sheet in Figure 15 rather than just making up the categories.**

# Qualitative surveys and interviews

In some geography fieldwork we want to investigate people's ideas, opinions, or perceptions. This is an example of **qualitative data** and we can collect this type of evidence through surveys, questionnaires, and interviews. We can use **bipolar Surveys** and **Likert Surveys** (pronounced lick it) to collect evidence about the strength of people's opinions. These surveys use **closed questions** – the people who take part have to choose the answer they think they most agree with.

**A closed question** is the type of question that needs a simple response: the person has to answer yes/no, or tick a box. Multiple choice tests use closed questions. This technique is good because it is quick and easy to analyse the results – you just count up the number of people who respond in each way.

## Bipolar surveys

**Bipolar Surveys** use pairs of statements that are opposite to one another. The people who take part in the survey have to decide where their opinion fits on the scale between these opposing statements.

A basic bipolar survey uses pairs of adjectives. Figure 16 gives an example that could be used to assess how people feel about one place. We could use this type of survey to investigate how different people perceive the same place. It could be a town, village, honeypot site or coastal landscape. For example:
- whether younger people are more positive about a place than older people
- whether visitors see more positives in a place than local people.

We can also use bipolar surveys to investigate opinions about a variety of different places. For example, to investigate how quality of life varies across a town or city. You don't even have to ask other people to do the survey – you and your classmates could fill it in. You would need to decide on a sampling technique to choose the places – perhaps conducting the survey every 200 metres along a transect that starts in the town centre and ends in the suburbs. If so, your bipolar survey could be one strand of evidence to investigate a hypothesis such as:

*Quality of life improves towards the suburbs*

**Figure 16** A simple bipolar survey uses pairs of opposite adjectives.

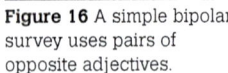
http://www.thesaurus.com/
Use the search engine to find synonyms (similar words) and antonyms (opposite words).

|  | +3 | +2 | +1 | 0 | -1 | -2 | -3 |  |
|---|---|---|---|---|---|---|---|---|
| Attractive |  |  |  |  |  |  |  | Unattractive |
| Quiet |  |  |  |  |  |  |  | Noisy |
| Peaceful |  |  |  |  |  |  |  | Calm |
| Safe |  |  |  |  |  |  |  | Unsafe |
| Dirty |  |  |  |  |  |  |  | Clean |
| Friendly |  |  |  |  |  |  |  | Unfriendly |

A slightly more sophisticated bipolar survey uses opposing descriptions instead of simple adjectives. An example is shown in Figure 17. If the descriptions are well written, it should be possible for different people to use this type of survey consistently – meaning that the results of your survey will be more reliable than the basic type.

**Figure 17** This bipolar survey could be used to test people's perception of a residential area.

| Positive | +2 | +1 | 0 | -1 | -2 | Negative |
|---|---|---|---|---|---|---|
| Housing well-maintained | | | | | | Housing in poor condition |
| Not much traffic | | | | | | Very difficult to cross the road |
| Wide pavements | | | | | | No pavements |
| Streets are well lit at night | | | | | | No street lighting |
| Area feels safe. There is no graffiti | | | | | | Area feels unsafe. There is a lot of graffiti and litter |

**Figure 18** A residential area in Grangetown, Cardiff.

## Dos and don'ts of bipolar surveys

**Do:**

✓ Make sure that the pairs of statements are actually opposites.

✓ Put the statements the right way around in the table – so that positive statements get a positive score.

**Don't:**

✗ Make your statements too complicated or try to assess two things at once in one pair of statements, for example, Quiet and attractive > > > > > Noisy and unattractive

## Activities

1 **Study Figure 16.**
   a) Identify two errors in this survey that would make it unreliable.
   b) Suggest how these errors should be corrected.
2 **Study Figure 17.**
   a) Identify two errors in this survey that would make it unreliable.
   b) Add two more pairs of bipolar statements that could be used in this fieldwork location.

| Strengths and limitations of bipolar surveys | |
|---|---|
| **Some strengths** | **Some limitations** |
| • Bipolar surveys are an easy way to collect evidence about people's opinions.<br>• The survey is quick so people don't complain about giving up too much time.<br>• The results are quick to process and analyse. Results from a large number of bipolar surveys are much easier to analyse than interviews and questionnaires that use open questions. | • Adjectives used in a bipolar survey may mean different things to different people. For example, in slang, the meaning of words like sick, bad, and wicked is the opposite of what they usually mean. If some people in your survey misunderstand the way you have used the adjectives, then your results will be unreliable.<br>• Some people always choose to score zero in the middle of the two adjectives. You can prevent this by deleting the option to score zero. |

# Likert Surveys and Questionnaires

As well as bipolar surveys, we can use Likert Surveys, questionnaires, and longer interviews to gather evidence about people's opinions.

## Questionnaires and interviews

Questionnaires and interviews allow people to talk openly and in detail about a geographical issue. In both of these types of survey the interviewer needs to prepare some questions so that the survey has a common structure each time it is used. This will allow you to compare the responses of different people and make the data more reliable. In your analysis you can then look for common patterns and similarities as well as differences between the responses.

A **questionnaire** uses a mixture of closed and **open questions**. The closed questions will allow you to collect a lot of data quite quickly. The open questions allow the interviewee to say what they want, without having to choose from a list of set responses. The advantage of using open questions is that people sometimes say things you didn't expect. This means that open questions may allow you to collect some new and surprising evidence. However, it can be time consuming to analyse the results of open questions so it's a good idea to only use a few open questions if you plan to survey a large number of people.

**Interviews** only use open questions so they can provide a wide variety of responses. It is a good idea to make an audio recording of each interview so you can listen to them again. That will help you to focus on the words people use and analyse any differences and similarities between what people say. See pages 58-61 to find out how to analyse text. To save time you may decide to interview only a small sample of people. The disadvantage of this is the views you record may not represent the views of everyone in the community.

## Dos and don'ts of questionnaires

**Do:**
- ✓ Be polite and courteous.
- ✓ Carry out a pilot survey (see pages 22-23) to identify any potential problems with the way your questions are worded.
- ✓ Use a mixture of closed and open questions.

**Don't:**
- ✗ Ask two things in one question, for example, 'Do you think that the new hotel would be good for local businesses and the environment?'.
- ✗ Ask people about their age or gender unless your aim is to see whether certain groups of people have a different point of view.

## Activities

1 **Study Figure 19.**
   a) Use this poster to design an enquiry question or hypothesis.
   b) Identify one more statement that could be added to the Likert Survey in Figure 20.
   c) Draft three open questions you could use to investigate the proposed new hotel in St Davids.

## Likert Surveys

In a **Likert survey** we ask people to what extent they agree, or disagree, with one or more statements. The statements could be about any geographical issue where different people might have strong points of view. We would use this type of survey to investigate whether some groups of people have a different point of view to another group. For example, we could use a Likert Survey to investigate whether people agree:

- with building houses on a greenfield site;
- that flood defences need improving/strengthening;
- that tourism is good for the local area;
- with the closure of a rural service such as the village post office.

It may be that younger people have a different point of view to older people, or that local residents tend to take one point of view while visitors take another. Different points of view, and the reasons for them, maybe the aim of your investigation. If so, you will need to design a data collection sheet that records this information. An example is shown in Figure 20. Then, when you analyse the results of your survey, sort the responses by type first (by age group, or by local/visitor) before tallying up the scores for agree/disagree.

**Figure 19** A poster in the street in St Davids, Pembroke. In 2017 some local people were protesting about plans to build a Premier Inn Hotel in the town.

**Figure 20** A Likert Survey that could be used to investigate opinions about the building of a new hotel.

Please tick the box that best describes you:

I am a visitor to St Davids ☐      I live in St Davids ☐

| | Strongly agree | Agree | Neutral | Disagree | Strongly disagree |
|---|---|---|---|---|---|
| The town would benefit from more tourists staying overnight | | | | | |
| The proposed hotel would create local jobs | | | | | |
| Other businesses in the town would benefit | | | | | |

# Designing data collection sheets for an Environmental Quality Index (EQI)

Urban fieldwork often involves the collection of environmental evidence. For example, you might want to collect data on:

- safety of the environment (quality of pavements, traffic calming, CCTV);
- pollution (noise, litter, graffiti);
- access (safe places to cross roads, cycle paths, ramps into shops);
- green spaces (amount of vegetation, how well they are maintained and used).

In order to collect this kind of data you will need to design an **Environmental Quality Index (EQI)** sheet. Precise criteria should be used in your sheet. These are statements that are used to quantify each feature of the environment being assessed. This will enable you to get consistent results if several people are collecting the data. Each criterion can then be numbered using, for example, a scale of 1 to 5. This means you can process the data later by calculating means and drawing graphs and maps. An example of this is shown in Figure 21.

It is a good idea to carry out pilot surveys to assess the effectiveness of your EQI sheet before you use it on the fieldtrip. This could be done near to the school, or even in the school grounds, so that the criteria can be discussed with your colleagues. You can then amend the EQI data collection sheet if necessary before the actual fieldtrip.

**Figure 21** Use precise criteria to assess each feature in an EQI survey.

Don't forget to add the following to your data collection sheet:

- a space to record the location of the survey;
- a space to record the time and date of the survey.

| Feature | Criteria | | | | |
|---------|----------|---|---|---|---|
| | 5 | 4 | 3 | 2 | 1 |
| **Vegetation** | Plenty of trees. 1+ tree per 20m. | Some trees. 1 tree per 20-39m. | Few trees. 1 tree per 40-79m. | Sporadic trees. 1 tree per 80-100m. | No trees or greenery. |
| | 5 | 4 | 3 | 2 | 1 |
| **Litter** | Occasional litter. One item every 50m+. | Hardly any litter. One item every 11-50m. | Some litter. One item every 6-10m. | Lots of litter. One item every 1-5m. | Abundant litter. More than one item per metre. |

## Using your phone to measure noise levels

You can use your smart phone to measure noise. You will need to download an App that measures decibels (dB) – there are several free ones available.

Normal conversation is about 65dB. The decibel scale is logarithmic. This means that 75dB is 10 times louder than 65dB and 85dB is 100 times louder than 65dB. Noise on a busy city street will be between 75dB and 85dB.

## Adding weightings to an EQI

It is possible that the features included in the survey may not have equal importance. For example, local residents may tell you that they are much more affected by traffic noise and litter than they are by graffiti. If so, you can increase the weighting of the features that are considered to be more important. Do this by multiplying them by a factor, as in Figure 22.

**Figure 22** Use weightings to reflect the view that some features of the environment are more important than others.

| Feature | Criteria | | | | |
|---------|----------|---|---|---|---|
| | 5 | 4 | 3 | 2 | 1 |
| **Vegetation** | Plenty of trees. 1+ tree per 20m. | Some trees. 1 tree per 20-39m. | Few trees. 1 tree per 40-79m. | Sporadic trees. 1 tree per 80-100m. | No trees or greenery. |
| **Litter** | 5x2=10 | 4x2=8 | 3x2=6 | 2x2=4 | 1x2=2 |
| | Occasional litter. One item every 50m+. | Hardly any litter. One item every 11-50m. | Some litter. One item every 6-10m. | Lots of litter. One item every 1-5m. | Abundant litter. More than one item per metre. |
| **Traffic** | 5x3=15 | 4x3=12 | 3x3=9 | 2x3=6 | 1x3=3 |
| | Occasional traffic. Less than 5 cars per minute. | Hardly any traffic. 5-9 vehicles per minute. | Some traffic. 10-15 vehicles per minute. All cars. | Lots of traffic. 16-20 vehicles per minute. Lorries and cars. | Traffic is a nuisance. More than 20 vehicles per minute. Lorries and cars. |

| Strengths and limitations of EQI surveys | |
|---|---|
| **Some strengths** | **Some limitations** |
| EQIs can include a wide range of environmental factors. They provide quantitative data which can easily be presented. Scores collected by different groups of students can be averaged to improve reliability. Precise criteria for each feature will increase reliability and comparability of results. If EQIs are collected along a transect, trends over distance can be analysed. If EQIs are collected at different places (in two dimensions) the numbers can be mapped and spatial patterns can be analysed. | Scores given may be affected by the time of day, week, or year. Scores for each feature will be inconsistent if groups of students who are using the EQI haven't agreed the criteria carefully first. |

## Activities

1 **Study Figure 21.**
   a) Explain why the scoring gives 5 marks for 'plenty of trees' and the same score for 'occasional litter'.
   b) Write a set of criteria for:
      i) Graffiti/vandalism
      ii) Condition of pavements
      iii) Access to shops/services
   c) Explain why graffiti/vandalism, condition of pavements and access to shops might be considered to be useful indicators of the quality of an urban environment.

2 **Study Figure 22.**
   a) Describe how you might use a pilot survey to decide how to set these weightings.
   b) Explain why weightings like these are useful.

# Collecting qualitative data - making images during fieldwork

You can collect images as evidence of your fieldwork by:

■ taking photographs;

■ making **field sketches**. These are simple outline drawings of the landscape;

■ creating **sketch maps**. These are simple maps of the area in which you do your fieldwork. They do not need to be to scale.

Each of these images can be labelled or annotated later when you are analysing your evidence (see page 32).

You might also take video during your fieldwork to collect evidence of, for example:

■ the complex way that people move through pedestrianised streets;

■ the ebb and flow of traffic during rush hour.

The evidence shown in photographs, videos, or sketches is a form of **qualitative data**.

## When should images be taken?

If you decide to make images during your fieldwork it is important that you make sure that each image serves a purpose. For example, you might:

■ use an image to provide an overview of the main features of the study area. Field sketches and sketch maps are particularly useful for this because you can concentrate on just showing the main features of the landscape that are relevant to your fieldwork;

■ take a photograph to illustrate the way that data has been sampled. You may be able to annotate this image to analyse the strengths and limitations of the data collection technique;

■ take a series of photographs to provide evidence of individual features that are relevant to your fieldwork. For example, you might want to take photographs of the features that people have identified as being important in a questionnaire or longer interview.

http://www.gridreferencefinder.com/ Use this simple website to find six figure grid references for your fieldwork locations and photos.

### Dos and don'ts of taking photos

**Do:**

✓ Provide a reference for the location of any photo that you take for your investigation. An OS grid reference is often best. See the web link above.

✓ Take photos in public spaces and respect others – some people do not want to be photographed.

✓ Always ask permission if you want to take a close up of someone, like the photo of Mr Pugh on page 90.

✓ Explain that your photos are only for your personal studies and that you won't be sharing them on social media.

**Don't:**

✗ Take photographs unless you can analyse them or use them somehow. You are not going to make a scrapbook.

✗ Stand in the way of other pedestrians or block pavements.

✗ Take photos on private property unless you have permission. This includes shopping centres.

✗ Take photos of people inside their own homes (through a window) even if you are in a public place because people have a right to privacy.

✗ Take photos of homeless people without their permission – be respectful.

# Why draw a sketch map of your fieldwork area?

It might be easy to download a map of your fieldwork area from a website. However, these maps are often cluttered and may be difficult to use. You might decide to use a sketch map instead because you can:

- leave out unnecessary details;
- just include the main features;
- label the features that are important to your fieldwork.

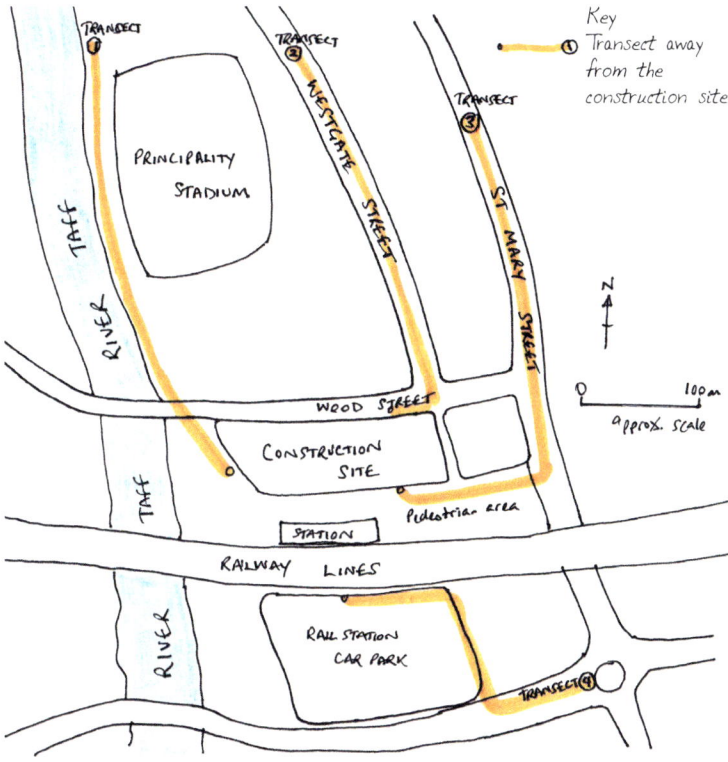

**Figure 23** A sketch map of a fieldwork location can focus on just the essential features. This example shows a potential fieldwork location in Central Cardiff.

**Figure 24** Apps on your smart phone can be used to record date, location, and altitude. This is the building site shown in the centre of Figure 23. An altitude app on the smart phone used the phone's camera and recorded altitude and location in the bottom left of the photo.

## Dos and don'ts of making sketch maps

**Do:**
- ✓ Give your map a heading, a north arrow and an approximate scale line.
- ✓ Make a key to match any colours or symbols on your map.

**Don't:**
- ✗ Add too many details and clutter the map.

## Activities

1 **Figure 23 was created at the postcode CF10 1EP.**
   a) Use a search engine to find a map of this postcode and its surroundings.
   b) Compare the search engine map to the sketch in Figure 23. What features have been left out of the sketch? Suggest why this makes the sketch more useful.

2 **Suggest why it is important to record time, location and altitude as evidence when making a fieldwork image.**

**Learning objectives**
- Why field sketches are useful.
- How to write annotations.

# Field sketches

Making a **field sketch**, like Figure 26, is another useful way to collect qualitative data. A field sketch should be used to record evidence about the most important features of a fieldwork location. Focus on drawing just the important geographical features. Leave out anything that isn't relevant. You should add notes to the sketch straight away while you are still in the field, for example:

- a grid reference to locate the sketch;
- labels for the main features;
- more detailed notes (annotations) that help to make sense of the features in your sketch.

**Figure 25** Photo of a river cliff on the River Onny in Shropshire.

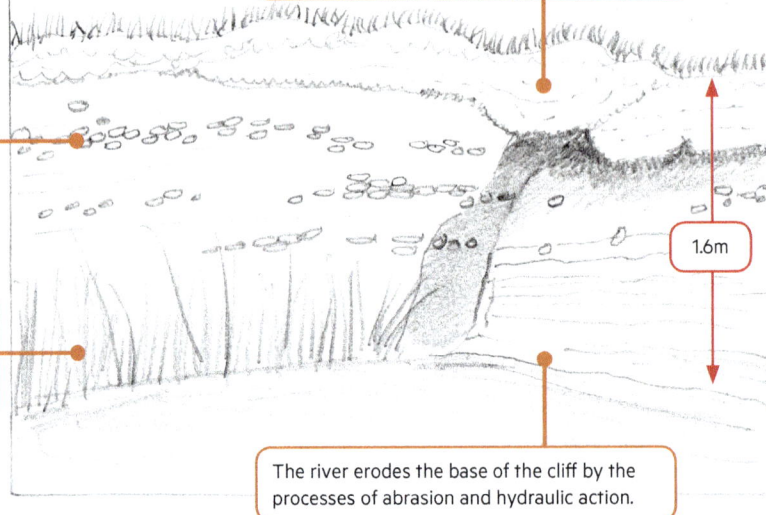

An overhang at the top of the river cliff has been formed by erosion at the base of the cliff.

Layers of sediment in the cliff is evidence of past deposition. The stones are a similar size and shape to those on the bed of the present-day river.

1.6m

The grass growing in soil that has slumped down when the river eroded the base of the cliff.

The river erodes the base of the cliff by the processes of abrasion and hydraulic action.

**Figure 26** Annotated field sketch of the river cliff at grid reference 344910.

## Dos and don'ts of field sketches

**Do:**
- ✓ Look at the feature carefully. If it's a small feature, like Figure 25, take measurements while you are there.
- ✓ Include a scale line.

**Don't:**
- ✗ Include any unnecessary detail or clutter.

# Annotation

**Annotation** means to add text that makes sense of an image. Annotation is more complex than labels which are simple descriptions. Annotations should be used to analyse key features of the image. The analysis should demonstrate your understanding of those features based on the evidence you collect during fieldwork.

You can annotate field sketches and photos of your fieldwork area. Figure 27 gives an example. The photo shows a range of hard engineering strategies that protect the low lying coastal landscape at Sea Palling in Norfolk. The annotations show understanding of how each strategy works. Notice that the annotations are numbered so that the reader is encouraged to read them in sequence. This helps tell the story given by the data.

**Figure 27** Use annotations to analyse the evidence in a photograph. Sea Palling, Norfolk.

4 The embankment protects the land from flooding during a storm surge when low air pressure causes sea levels to rise.

3 Sea wall. The concrete sea wall is curved to reflect the energy of the largest waves.

2 Revetment. The concrete steps break up the force of the waves.

1 Rock armour. These boulders absorb wave energy. They also prevent backwash from the sea wall and revetment from eroding sand at the foot of the steps.

| Strengths and limitations of annotation | |
|---|---|
| **Some strengths** | **Some limitations** |
| Annotation should help you to pick out and analyse the key features of an image quickly and precisely. Annotation can be used to analyse a wide variety of images including photographs, field sketches, and sketch maps. | You may have a limited amount of space for your annotation so choose your words carefully. |

## Activities

1 **Study the river cliff in Figure 25.**
   a) Suggest an enquiry question or hypothesis that could be investigated in this location.
   b) Design a sampling strategy that would provide further evidence about how the stones in the river cliff were deposited.

**Learning objectives**
- Being aware of secondary sources that can be useful in the field.

# Using secondary sources to help collect data

A lot of data is available in books and on websites. This is **secondary data** and it can be used at several stages of the enquiry process as shown in Figure 28.

**Figure 28** How secondary data may be used in fieldwork

| Stage of the enquiry process | How to use secondary data |
|---|---|
| Planning | Photographs of your fieldwork location from the internet can be used at the planning stage to help you create a hypothesis or pose an enquiry question. |
| Data collection: Sampling procedures | Tables of data can be used to help you select a sampling strategy. An example is shown on page 37. |
| Data collection: Collecting qualitative data | Blogs can be used to collect evidence of people's viewpoints. |
| Data collection: Collecting quantitative data | Many websites contain huge quantities of data. The use of these **big data** sets is described on pages 36-37. |
| Data presentation | Tables of data from websites can be used to draw maps and graphs. |
| Analysis | Websites may include graphs and maps that are relevant to the aims of your fieldwork. Select some and annotate them to help your analysis. |

## Dos and don'ts of using secondary data

**Do:**
- ✔ Select secondary data carefully. Only present and analyse data that will be useful to answer your aim.
- ✔ Acknowledge where your evidence has come from. Do this by giving the URL of the webpage where you found the data. Then you will be able to find the source again if you need it.
- ✔ Consider the reliability of the source. Can the evidence be trusted?

**Don't:**
- ✘ Copy photographs or other evidence from the internet into your portfolio if you don't intend to process or analyse it. You can collect too much stuff which will make your portfolio difficult to work with.

## Using your smart phone during fieldwork

Some apps, like the noise meter app described on page 28, turn your smart phone into a measuring tool for collecting primary data. You can also use your smart phone to provide secondary data that may be useful to you while you are working in the field. There are several apps that you might be able to use. Figure 29 shows four of them and suggests how they might be used during fieldwork.

## Activities

1. Students wanted to investigate a woodland growing on a hillside. Their enquiry question was *'Are species of tree affected by altitude or angle of slope?'*. The most common trees in the woodland were oak, birch, ash, hawthorn, and holly.
   a) Select and justify a sampling strategy for this enquiry.
   b) Select up to four apps that could be used and explain why you have chosen them.
   c) Design a data collection sheet for use with this enquiry.

Use a plant identification app to help you recognise plants. This might be useful if your enquiry investigated the effects of trampling in an ecosystem such as a sand dune or an urban park.

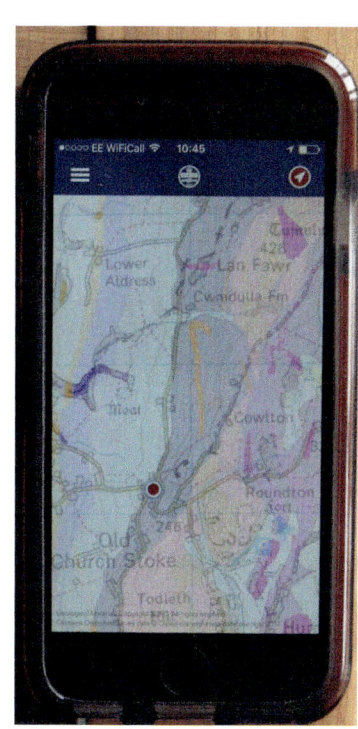

The iGeology app uses your phone's location to display a geological map of your location. You could use it to see whether local rocks are permeable or impermeable. This might help if your enquiry involves an investigation of flows of water through a drainage basin.

Use an app like this to measure distances, for example, when collecting data along a longer transect (or long profile) through a large town or across a rural area.

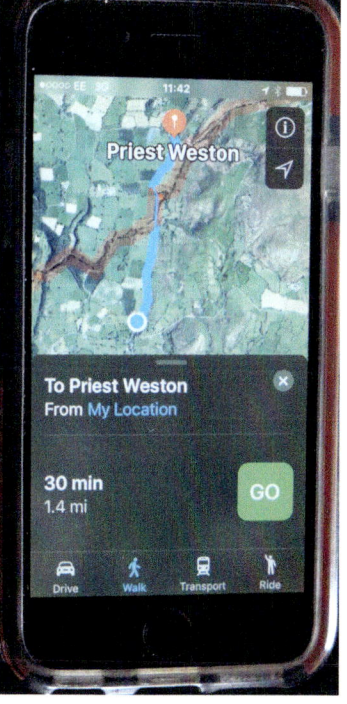

There are several apps that will display your elevation above sea level. This could be used to record your height if your transect went up and down a slope. It might also be useful if you were investigating whether land is at risk of flooding. This app will access your phone's camera so you can take images with a record of the time, location and elevation, as in Figure 24 on page 31.

**Figure 29** Smart phone apps that you might use during your fieldwork.

https://www.riverlevels.uk/ This website publishes discharge data for rivers in the UK. You can use it to see how the discharge of a river changes with time. It is too dangerous to collect river velocity data when the river is in flood, so this website is particularly useful for seeing how river discharge changes after heavy rainfall.

https://www.nomisweb.co.uk/census/2011 The official UK National Census website.

# Big data

Some websites publish very large databases containing huge amounts of evidence – sometimes known as **big data**. These websites provide secondary data that may be of use to your enquiry. For example, you may be able to find a connection between primary data you have collected and secondary data published on a website.

- House prices (on a website) might be explained by variations in quality of the urban environment you have found using an EQI.
- Daily rainfall for the last month (on a website) might help you explain patterns of discharge you have found in a river and its tributary.

## The National Census

Sometimes it can be difficult to answer the aims of your enquiry if you are only using primary data. For example, you may need to know such things as:

- the age structure of the population;
- the main occupations that people have;
- how many cars are owned by each household;
- the level of education that people have achieved.

You might be able to find this data through primary data collection by using a questionnaire. However, this would be very time consuming. Besides, people are sometimes reluctant to talk about their age, their education or their jobs so collecting this as primary data would be tricky. Thankfully, the National Census collects and stores this kind of evidence. It contains information about every household in the UK and includes data about the UK population's age, health, occupations, and education, amongst many other things. The census does have one major limitation: it is a snap-shot in time, recording data every 10 years. Your fieldwork could be located somewhere that has changed significantly since the last census. For example, an enquiry that investigates the sustainability of a new housing estate could not be supported by secondary data from the census if it was built after the census date.

**Figure 30** Examples of enquiry questions that would need some secondary data.

1. Are house prices higher closer to the park?
2. Is it healthier to live in the suburbs or in the city centre?
3. Are fewer crimes recorded in areas that have more CCTV cameras?
4. Are the residents of sustainable communities mainly in professional jobs?
5. Is car ownership higher in suburbs than in the city centre?

https://www.streetcheck.co.uk/ This website combines UK National Census with other big data sets such as data on crime and house prices. It can be used when investigating concepts such as inequality or sustainability.

The street check website takes some of the data from the UK National Census and combines it with other big data sets such as data on crime and house prices. It can be searched using place names or postcodes. It provides numbers that you can process. It also presents some data in the form of graphs. You can search the database by postcode. A typical postcode unit (like CF5 2YX) has 200-300 people living in it. A postcode district (like CF5) can have between 8,000 and 85,000 people living in it, depending on whether the area covered by the postcode is rural or urban. It would be SMART to focus your fieldwork enquiry on a small number of postcode units rather than a postcode district which would be too large.

| Strengths and limitations of using census data | |
|---|---|
| **Some strengths** | **Some limitations** |
| The census covers the whole of the UK. You can use it to compare data in one neighbourhood to another so you can analyse spatial patterns. The census is repeated every 10 years so you can find evidence of how a place is changing (for example, by comparing data from 2011 with 2001). | Some places change very quickly so the census may not accurately reflect what a place is like on the day of your fieldtrip. |

# Using census data to select a sample

You can use big data to research your fieldwork location before you visit. For example, you could use the census to decide on a suitable sample of local residents. A suitable **sampling strategy** for a questionnaire would be a **stratified** sample – where the number of people in each age group that you interview is in proportion to the actual population. You would do this by completing the following steps.

**Step One** Use a website with census data. Select the **ward** and find the population structure data. Figure 31 shows the age structure of Borth, a small seaside town in West Wales.

**Step Two** Calculate the percentage of people in each age category. For example, to calculate the percentage of people who are over 65 in Figure 31:

Divide 460 by 2,070 (the total) and multiply the answer by 100 = 22.22

This means that, if you questioned 100 people in total, you should interview 22 who are over 65.

**Step Three** Calculate the number of people to interview if your total sample size is less than 100. So, if you only want to interview 50 people in total, you multiply the answers from Step Two by 0.5 (because 50 is half of 100).

| Age of residents | Number of residents in each age group |
|---|---|
| Over 65 | 460 |
| 30–64 | 1000 |
| Under 30 | 610 |
| Total population | 2070 |

**Figure 31** The age structure of Borth, West Wales. National Census (2011).

## Activities

1  **Use the data in Figure 31.**
   a) Calculate how many people would be interviewed in each age category in a stratified sample of 100.
   b) Calculate how many people would be interviewed in each age category in a stratified sample of 60.
2  **Study Figure 30. It gives examples of enquiry questions which rely on at least some secondary data.**
   Use the street check website to identify at least one type of secondary data that you could use to help you answer each of these enquiry questions.

# Geographic Information Systems (GIS)

Lots of websites use interactive maps to show geographical data. They are known as **Geographical Information Systems (GIS)**. These maps are built up from a number of different layers. There is usually a simple base map which shows roads and place names. Then, on top of the base map, there are various layers of data that you can toggle on or off so that you can choose which layer to view. The data is often shown using the colour shading technique used in **choropleth maps**. Figure 32 shows one example. This map shows patterns of deprivation (a measure of inequality). Areas coloured blue/green have a better standard of living than the UK average whereas areas coloured red/orange have a lower standard of living than the UK average.

## Using GIS to design a sampling strategy

We can use GIS before the fieldtrip while we are planning our aims and sampling strategies. For example, we could use Figure 32 to plan a transect (sampling along a line) from Birmingham's suburbs towards the city centre. We could use the GIS to check that our proposed transect goes through districts with different levels of deprivation.

## Using GIS to investigate socio-economic data

We can use GIS as a source of secondary data to support the **primary data** collected during the fieldtrip. Several GIS websites display data to do with population, crime, health, house prices, or deprivation. This data is about people and the economy so it is sometimes described as socio-economic data. Figure 33 shows a screenshot from a useful GIS where you can explore crime data. Notice that the data is shown in the form of proportional circles. As you zoom in, the circles break up so you can pin-point where the crimes were reported more accurately.

You might be able to use a combination of primary and secondary data (from a GIS) to help investigate the connection between two sets of data. For example, we could investigate whether:

- traffic is busier (primary data) in areas of higher population density (GIS data);
- there are more CCTV cameras (primary data) in areas which have higher crime rates (GIS data);
- house prices are higher (GIS data) in neighbourhoods that have more green spaces (primary data).

| Strengths and limitations of using GIS as a source of data | |
|---|---|
| **Some strengths** | **Some limitations** |
| • The data can be trusted because it has been collected by reliable sources.<br>• Complex data has been presented using maps that are easy to understand so you can quickly see spatial patterns. | • GIS uses secondary data that might not be absolutely up to date.<br>• GIS maps use data from areas (or districts) that contain similar numbers of people. These areas can be quite large in rural areas so it is not possible to see data for smaller, local districts such as individual villages. |

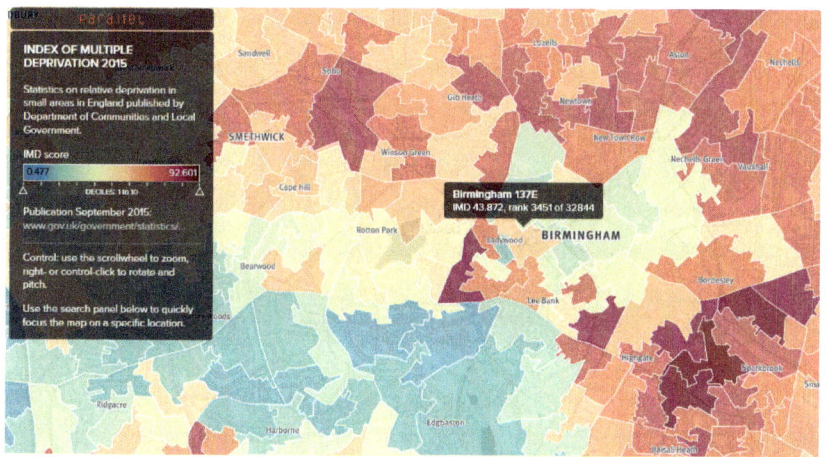

https://parallel.co.uk/ This GIS maps data such as population density, deprivation, fuel poverty, and air pollution.

**Figure 32** A screenshot from parallel GIS. This GIS uses colour shading to show an index of deprivation. The data for each area pops up as the cursor moves over the map.

**Figure 33** A screenshot from the police.uk GIS. This shot shows reported crime in Derby.

https://www.police.uk/ This GIS uses an interactive map to represent recent reports of crime.

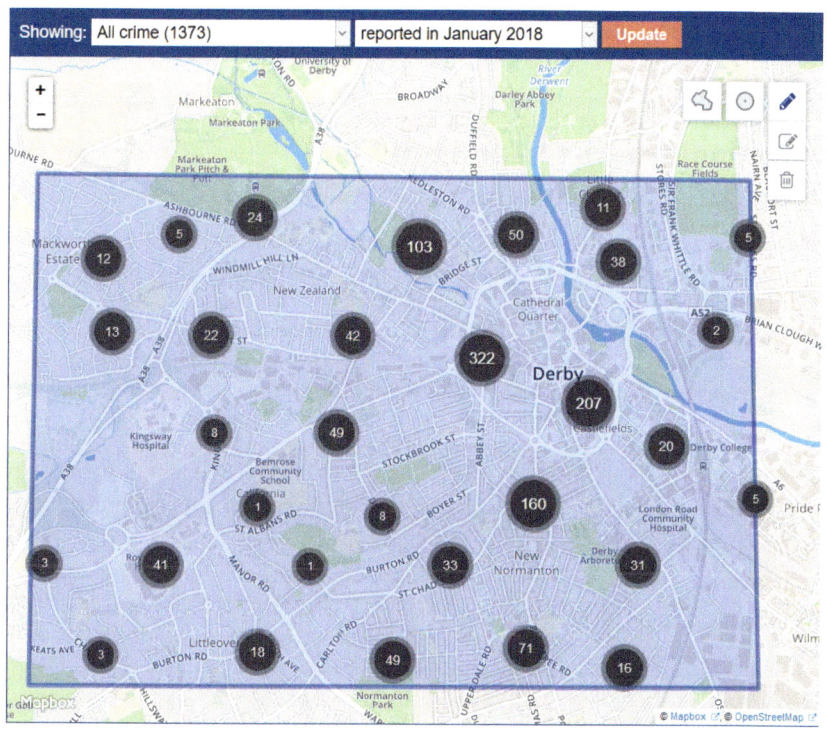

## Activities

1 Suggest one strength and one limitation in using GIS to plan your enquiry, or design a sampling strategy.

2 Suggest three different enquiry questions, or hypotheses, that could be investigated using a combination of primary data and secondary data from one of the GIS maps shown on this page.

# Chapter 3

# Processing and presenting evidence

## Learning objectives

- Knowing when and how to use mean, median, and mode.

**Figure 1** A student has collected data on wind speeds at three locations on a hillside. The data is shown in the table (values in metres per second).

# What do we mean by average?

Processing your data will allow you to see what the evidence is telling you. One useful processing technique is to find the average. This is a measure of **central tendency**, in other words, the average tells you something about the central point in your dataset. There are three different ways to express an average. These are mean, median, and mode.

## Mean

The arithmetic **mean** value of a dataset is found by:

**Step One** Adding up all the data to find the total.

**Step Two** Dividing the total value by the number of items of data.

Geographers often use the mean to find the average for **discrete** or **continuous data**. It is especially useful if data has been recorded several times and varies a little each time. For example, you might take five wind speed readings at each location (as in Figure 1) and find the mean. The wind may be gusty so each reading is different. By calculating the mean you can average out these variations. Small variations can also be caused by experimental error. For example, you might measure the time it takes for a dog biscuit to float downstream over a length of 10 metres. You might repeat the experiment 5 times and find each reading is slightly different because you didn't stop the stop watch promptly each time. Then you can take the mean to find an average time that reduces the effect of this experimental error.

| Site A – lower slope | 4.2 | 3.9 | 3.7 | 3.8 | 4.1 |
|---|---|---|---|---|---|
| Site B – middle slope | 3.9 | 4.2 | 4.4 | 4.2 | 4.0 |
| Site C – upper slope | 4.8 | 5.1 | 4.7 | 4.8 | 5.2 |

## Median

The **median** is the middle value when all your data is arranged in rank order. Median is a useful measure to use if you have some extreme values in your dataset. The median is unaffected by these extremes so will still give you a fair measure of the central point in your data. To find the median value:

**Step One** Put the data in rank order.

**Step Two** The median is the middle value in an odd number of items of data. In Figure 2, the median value in Group A is 5. If you have an even number of items in your dataset, the median is the half way point (mean) between the two middle values in your dataset. In Figure 2, to find the median value of Group B, add 6 to 12. Then divide the answer by 2.

**Figure 2** A student has collected data on the length of commute for three groups of workers in an office in Cardiff. The data is shown in the table (values in miles).

| Group A | 2 | 2 | 3 | 5 | 5 | 5 | 12 | 23 | 26 | |
|---|---|---|---|---|---|---|---|---|---|---|
| Group B | 1 | 2 | 4 | 4 | 6 | 12 | 18 | 22 | 56 | 103 |
| Group C | 1 | 1 | 2 | 3 | 4 | 7 | 8 | 12 | 15 | 43 |

# Mode

The **mode** is the most frequently occurring value in your dataset. Mode is a useful measure of discrete data in a geographical enquiry. For example, you could use mode to describe the average value collected in a **bipolar survey**, as in Figure 3.

**Step One** Tally all the times someone chose a particular value in the bipolar survey.

**Step Two** The mode is the value that has been chosen most frequently.

| | 2 | 1 | 0 | -1 | -2 | |
|---|---|---|---|---|---|---|
| **Pavements are in good condition** | 2 | 3 | 5 | 18 | 12 | **Pavements are in poor condition** |
| **Cars never park on pavements** | 5 | 10 | 8 | 6 | 1 | **Cars often park on pavements** |

Figure 3 A bipolar survey was used as part of an enquiry about reducing risk for pedestrians and cyclists in a busy town. 30 people were surveyed. This table shows the total number of people who chose each value on the scale +2 to -2.

Mode can also be used to find a central point in grouped data. For example, imagine a beach survey in which pebbles are measured and put into categories by size, as in Figure 4. The **modal class** is the category that has most pebbles in it.

| Pebble size (mm) | Number of pebbles |
|---|---|
| Less than 50 | 32 |
| 50-60 | 28 |
| 60-70 | 15 |
| 71-90 | 15 |
| 91-120 | 10 |

Figure 4 A student measured the length of 100 pebbles at one location on a beach. The data is shown in the table (values in millimetres).

| Strengths and limitations of using averages | |
|---|---|
| **Some strengths** | **Some limitations** |
| Mean will give a number on a continuous scale. This is useful if you want to process the data further, for example, by adding the mean value to a scatter graph (see pages 63-65).<br>Mode is always a data value and median will be a data value if you have an odd number of items of data. This is an advantage if you want to present findings that make common sense. For example, in a survey of car ownership, you may find that the mean number of cars owned per household is 1.86! It would be more sensible to report that the mode was 2 cars per household. | Mean can fail to represent the central point fairly if the dataset includes some extreme values or a very wide range of values. Median is a better measure if your data includes extremes. |

## Activities

1  **Calculate the mean wind speed at each location in Figure 1.**
2  **Study Figure 2.**
    a) Calculate the mean, median and mode for each group of commuters.
    b) Explain why the mean value does not fairly represent the central point of this data.
3  **Study Figure 3. What is the mode for each statement? What does this tell you about this survey?**
4  **Study Figure 4.**
    a) Which size of pebbles is the modal class?
    b) Identify two faults in the data collection sheet.

# Measures of dispersion

Dispersion is a description of how your values are spread (or dispersed) through a dataset. The simplest measure of dispersion is **range**. This is the difference between the highest and lowest value in your dataset. This is useful because the range of one dataset may be very different to the range of another, even if the average value of the two datasets is similar. For example, in Figure 5, quadrats have been used to sample pebbles at two different locations on a beach. The table shows the length of 15 pebbles sampled from each quadrat. The range in values is quite different even though the median is the same. This suggests that wave action has moved pebbles up and down the beach and has sorted them by size – leaving pebbles with a very narrow range of sizes further up the beach.

**Figure 5** Pebble sizes at two locations on Vik beach, Iceland.

| Length of pebbles (mm) | |
|---|---|
| **Quadrat 1** | **Quadrat 2** |
| 95 | 25 |
| 60 | 22 |
| 35 | 20 |
| 30 | 20 |
| 20 | 18 |
| 18 | 15 |
| 14 | 14 |
| 14 | 14 |
| 11 | 12 |
| 10 | 11 |
| 9 | 10 |
| 9 | 9 |
| 8 | 9 |
| 8 | 7 |
| 7 | 8 |

**Figure 5.1** Quadrat 1. Pebbles close to the water's edge.

**Figure 5.2** Quadrat 2. Pebbles 5 metres from the water's edge.

## Interquartile range

The **interquartile range** is another way of measuring the spread of values in a dataset. The interquartile range is the difference between values that are one quarter and three quarters through the dataset so it excludes the largest and smallest values. It is used when the overall spread of values is rather large. To find the interquartile range follow these steps.

**Step One** Put the values into rank order and identify the median value. In Figure 5 it is in the green cell.

**Step Two** Identify the lower and upper interquartiles. The lower quartile is the mid-value between the lowest value in the dataset and the median. It is in the orange cell in Figure 5. The upper quartile is the mid-value between the median and the highest value and the dataset. It is in the yellow cell in Figure 5.

**Step Three** Calculate the interquartile range by subtracting the value of the lower quartile from the value of the upper quartile.

# Dispersion graphs

The range of values in a dataset can be presented visually using a **dispersion graph**, like Figure 6.2. Figure 6.2 shows the range in house prices in Borth, a seaside town. To make a dispersion graph follow these steps.

**Step One** Set out your categories of data on the horizontal **axis**. In Figure 6.2, these are different locations within Borth.

**Step Two** Mark the vertical axis with a range of values for your data. In Figure 6.2, this is house prices. The highest value is £380,000 so the axis is marked at £50,000 intervals up to £400,000.

**Step Three** Plot each value in your dataset as a small vertical cross. In this case, each cross represents the value of one house that is for sale in Borth.

**Figure 6.1** Borth in West Wales.

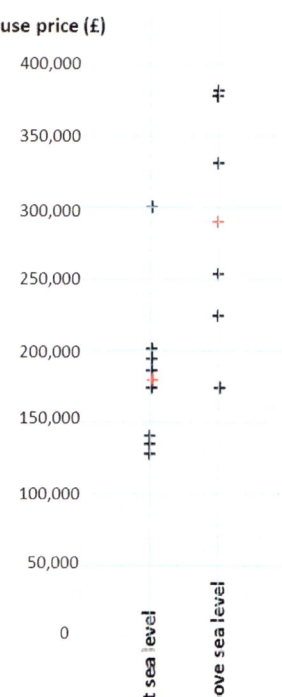

**Figure 6.2** Dispersion graph showing range of house prices in Borth (2017).

## Activities

1 **Study Figure 5.**
  a) For each quadrat, calculate the:
    i) range    ii) median    iii) interquartile range
  b) What conclusions can you draw from these results?
2 **Study Figure 6.2.**
  a) What do the red crosses on this graph represent?
  b) Use the graph to calculate the interquartile range in house prices for these two locations.

**Learning objectives**
- How to choose the right kind of graph for discrete and continuous data.
- How to present data using a line graph.

# Types of graph

Graphs are a way of presenting your data in a visual form. They make it easier to see patterns or trends within your data. They also make it possible to compare data easily.

There are lots of different types of graph you can draw but choosing which graph to draw is not just about personal taste. There are some rules about which graphs you can use.

Two of the most common graphs are:
- **Line graphs** which are used to present continuous data and can be used for discrete data.
- **Bar charts** (or column graphs) which are used to present discrete data and must **not** be used for continuous data.

Bar charts and line graphs each have two axes. The horizontal **axis** is known as the x-axis and the vertical axis is the y-axis. The place where the two axes meet is called the **origin** of the graph.

## Line graphs

**Line graphs** are used to present continuous data. **Continuous data** is data that is measurable and the result may be recorded to one or more decimal places. Distance, height, slope angle, sound level, velocity (for example, flow of a river or wind speed) and time are all examples of continuous data collected in fieldwork. Figures 9 and 10 each show continuous data displayed as line graphs.

**Discrete data** in one category, such as the amount of traffic, can be shown as a line graph if the x-axis shows time (for example, in hours, days, or years). Figure 7 shows an example. However, you should never present separate categories of data, such as different shop types, joined by a line.

The x-axis of a line graph can be used to show distance along a transect. For example, you can present the slopes of a beach profile as a line graph (see pages 20-21). Figure 10 shows the height of plants growing on a transect that cuts at right angles across a footpath. Cross sections, such as the shape of a river channel, can also be presented as a line graph.

| Time | Pedestrians |
|------|-------------|
| 8:00 | 36 |
| 8:30 | 78 |
| 9:00 | 69 |
| 9:30 | 43 |
| 10:00 | 44 |
| 10:30 | 53 |
| 11:00 | 57 |

**Figure 7: Pedestrians in Cardiff** Students collected data on pedestrian flows in a busy shopping street. They counted the number of pedestrians who passed by in three minutes every half an hour.

**Activities**

1 Discuss why it is important to start a line graph on the y-axis.
2 Suggest one strength and one limitation of using the y-axis to represent height in Figure 9.

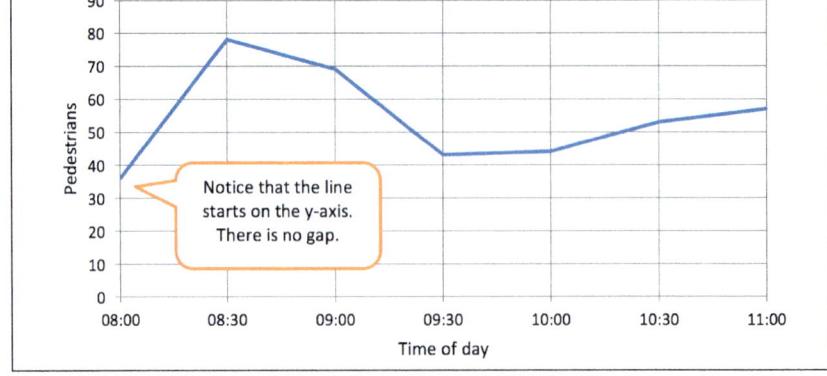

Notice that the line starts on the y-axis. There is no gap.

**Figure 8** Students collected data on flows in this rural honeypot site. They set up one transect up the west facing slope, from top to bottom. They collected wind speed data. They also set up a short transect across the footpath to see how plants were affected by trampling. Shropshire Hills, Area of Outstanding Natural Beauty (AONB).

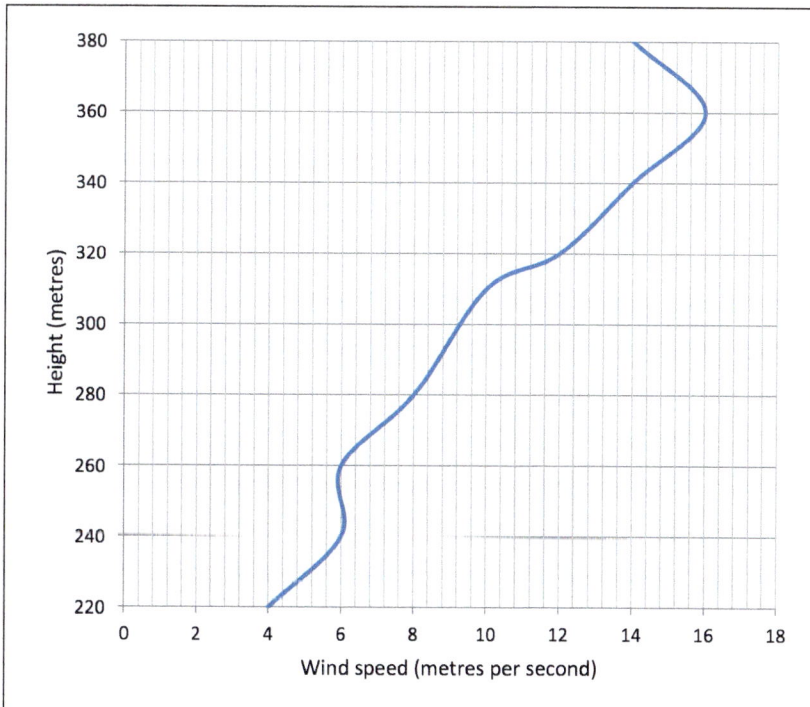

**Figure 9: Wind speeds**
Line graph of wind speeds. Notice that the graph shows height of the hill on the y-axis.

**Figure 10: Plant height**
Cross section across the footpath showing plant height

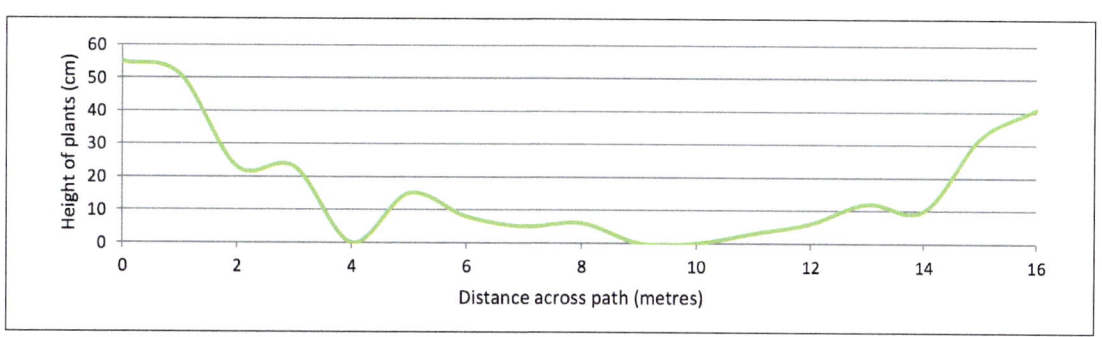

**Learning objectives**
- How to present data using a bar chart.

# Bar charts

You should choose to use a bar chart if you have **discrete data**. Discrete data is data that can be counted and the result is a whole number. The number of pedestrians in a shopping area and the number of cars passing in 5 minutes are both examples of discrete data. The number of items in a category, for example, the number of pebbles of a particular shape, is another example of discrete data. If the bars are drawn vertically as columns, then the x-axis is used for the categories of data and the y-axis is used for the values. For example, traffic can be shown as vertical bars with different categories of vehicle on the x-axis. Bar charts clearly show the category with the largest and smallest values so are useful for comparing your categories.

## Hints and tips for creating bar charts

- It is possible to draw horizontal rather than vertical bars. There is no rule to decide whether your bars should be vertical or horizontal but if your category names are very long then it is easier to label horizontal bars.
- Some categories have a logical sequence and should be shown in the correct order, like Figure 13. However, if your categories have no particular sequence then you can put your data in rank order by value from largest to smallest – making comparison easier. An example is shown in Figure 14.
- Decrease and increase can be shown by using positive and negative numbers on either the x-axis (like Figure 15) or the y-axis (if you want a column graph).

**Figure 11** Students collected data in this district shopping centre in Cardiff.

**Figure 12** Students recorded shop type on each side of the street. They compared their results to images on Google Street View in 2008 (secondary data).

| Shops in Cowbridge Road East | 2008 | 2016 |
|---|---|---|
| Food shops | 2 | 6 |
| Newsagents | 1 | 1 |
| Cafes/pubs/take aways | 20 | 26 |
| Estate agents | 2 | 3 |
| Health care/chemists | 6 | 2 |
| Furniture / home ware | 5 | 3 |
| Clothes | 5 | 4 |
| Electronics | 1 | 1 |
| Hair salon/beauty | 2 | 2 |
| Charity shops / betting / amusements | 10 | 18 |
| Vacant shops | 9 | 2 |

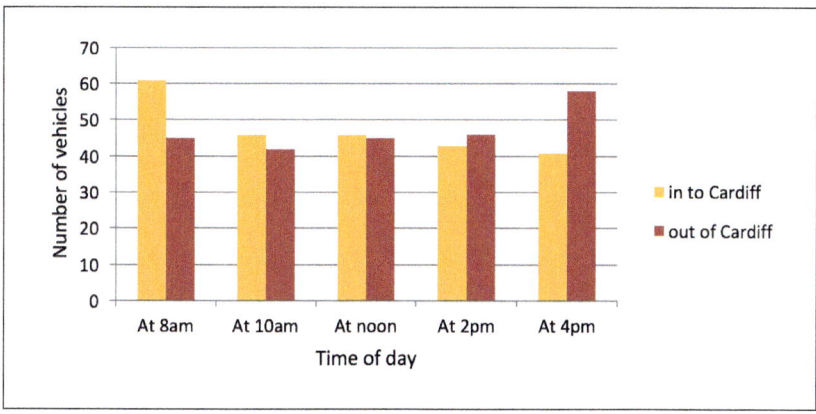

**Figure 13** Traffic data was collected in the street shown in Figure 11. Vehicles were recorded travelling in each direction for 3 minutes. The results are presented in a vertical bar (or column) chart.

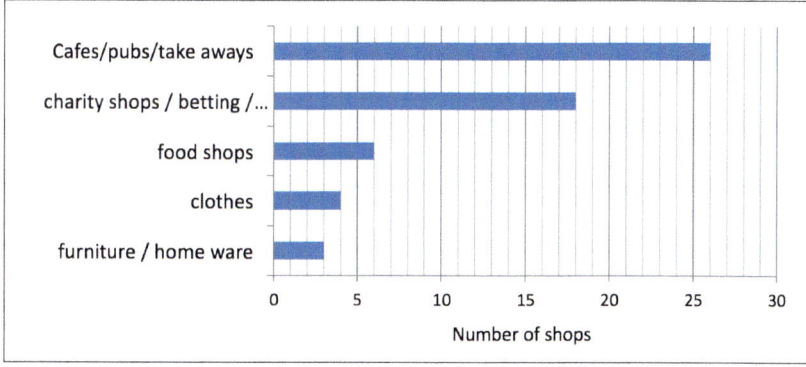

**Figure 14** The five most common shop types in 2016 are presented in this bar chart.

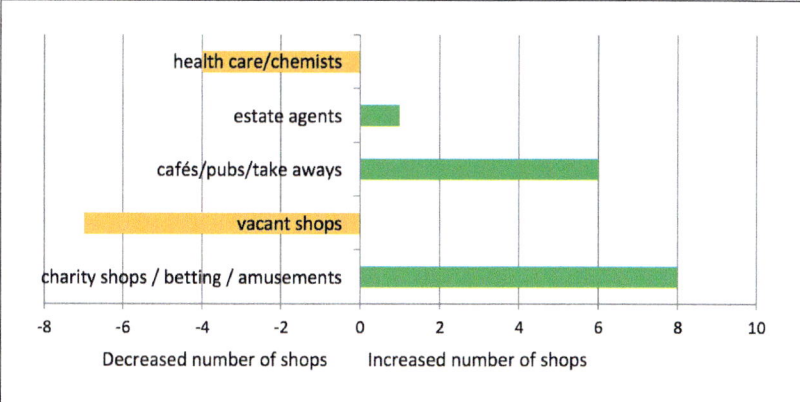

**Figure 15** The change in the number of shops in selected categories between 2008 and 2016 is shown in a bar chart that has an x-axis with both positive and negative numbers.

## Activities

1 **Study Figure 13.**
   a) Explain why a bar chart has been selected to present this data.
   b) What conclusions can you reach from this data?
2 **Use the data in Figure 12 to practise drawing:**
   a) a horizontal bar chart with the data in rank order;
   b) a vertical bar chart showing the change from 2008 to 2016.

# Pie charts and divided bars

**Pie charts** and **divided bars** are a useful way to present how a whole set of data can be divided up into different parts. Imagine you interviewed 100 people: 15 of them were retired and 12 were students. You could represent this data as a pictogram as shown in Figure 16. However, pictograms like this are time consuming to draw. A simpler and quicker method to show different parts of the whole set of data is to represent each part as a slice of a pie (a sector of a circle).

**Figure 16** Pictogram representing the number of retired people and students within a survey of 100 people.

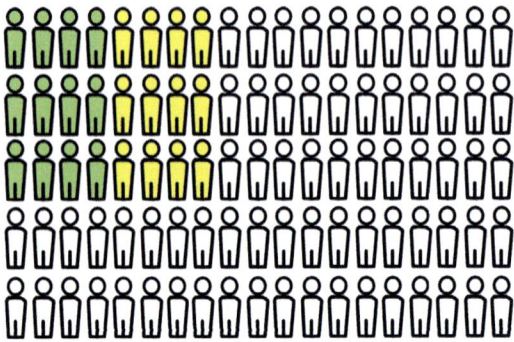

## Calculating percentages

To draw a pie chart or divided bar you will need to turn your raw data into percentages. Figure 16 is based on a questionnaire survey of 100 people so the maths is easy – each person is one per cent of the whole. But most data isn't quite as straightforward as this. Look at Figure 17. It shows data from the 2011 census for **wards** of Newcastle-under-Lyme in Staffordshire. It's an example of raw data – data that hasn't been processed. Notice that each ward has a different total number of residents. Loggerheads has most retired people. It also has more residents than other wards so it's difficult to know whether it has more or fewer retired people than other wards. To make comparisons between wards easier we need to convert this raw data into percentages.

**Figure 17** Census data (2011) for selected wards of Newcastle-under-Lyme, Staffordshire.

| Location | Ward | Residents | Retired |
|---|---|---|---|
| Inner urban | Town | 5,063 | 467 |
| | Knutton & Silverdale | 4,313 | 439 |
| | Cross Heath | 5,887 | 622 |
| Suburban | Westlands | 5,659 | 792 |
| Rural-urban fringe | Keele | 4,129 | 93 |
| | Loggerheads | 6,948 | 1,068 |
| Newcastle | | 123,871 | 14,830 |

To calculate the percentage of the population of Loggerheads ward that is retired you need to follow these steps.

**Step One** Divide the number of people in Loggerheads who are retired by the total number of residents in that ward. So, divide 1,068 by 6,948.

**Step Two** Multiply the answer (from Step One) by 100. So, multiply 0.1537 by 100. The answer is 15.37. This is the percentage of all residents in Loggerheads who are retired.

**Step Three** You can round the number up or down to the nearest whole number. If the number ends .49 or less, round it down. If it ends .50 or more, round it up. So, 15.37 would round down to 15%.

## Activities

1 **Study Figure 17.**
   a) Calculate the percentage of people who are retired in each ward of Newcastle-under-Lyme.
   b) Calculate the percentage of people who are retired in the whole of Newcastle-under-Lyme.
   c) Identify one anomaly in the data. Use the internet to research the anomalous ward and explain it.
   d) What conclusion can you reach about the data in Figure 17?

# Using pie charts and divided bars

To draw a pie chart by hand you will need a pie chart scale or a protractor. A complete pie represents 100% and a circle is divided by 360 degrees. So, to draw a pie chart using a protractor follow these steps.

**Step One** Calculate your percentages and check they add up to 100%.

**Step Two** Multiply each percentage by 3.6. For example, to make a sector of a pie for retired people in Loggerheads, multiply 15 by 3.6. The answer is 54. This is the number of degrees for the sector of the pie.

**Step Three** Draw a line from the centre of the circle to the top (at twelve o'clock if it were a clock face). Then measure around 54 degrees in a clockwise direction and draw a line from the centre to the edge of the circle. This is your first sector. Colour and label it.

Divided bars are even easier to draw and read. Simply draw a bar that is exactly 100mm long so that each 1mm length of bar equals 1 per cent.

Pie charts and divided bars are useful for presenting data where only a few 'parts' make up the whole. Figure 18 shows how effective they can be for making comparisons – in this case, between two different wards. Pie charts are much less effective if they have a large number of sectors.

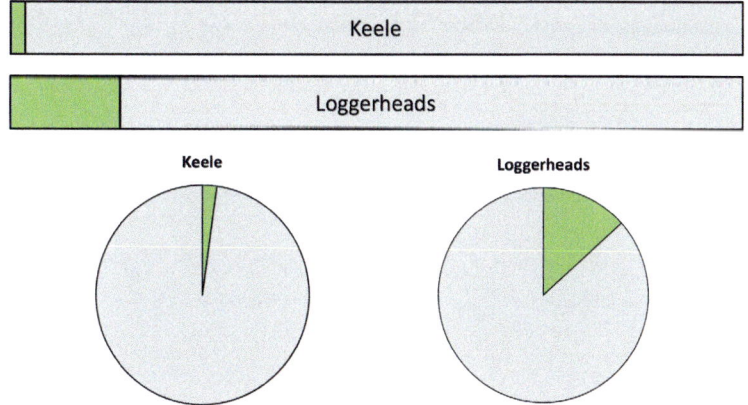

**Figure 18** These pie charts and divided bars represent the data for Loggerheads and Keele.

| Strengths and limitations of pie charts and divided bars | |
|---|---|
| **Some strengths** | **Some limitations** |
| • Pairs of pie charts, each with 2-5 sectors (or slices), are easy to compare.<br>• Pie charts can be located on a map to show spatial patterns (see page 91).<br>• Pies and divided bars simplify complex raw data into simple percentages so they are great ways to show differences and similarities between data when the raw data is of different sizes. | • Some people find pie charts difficult to read. Remember that a sector that is a quarter of the circle is 25% and a sector that is half the circle is 50%.<br>• Don't use a pie chart if you have too many sectors because people find it hard to judge the size of a small slice and lots of different coloured slices means your key will be complex. |

- How to avoid common mistakes when drawing graphs.

# Graphs: getting the details right

It is important to get the details right so that people who view your graph can fully understand what it is telling them about the data. A good graph, selected carefully to make sure it is appropriate for the data (**discrete** or **continuous**), will have:

- a Title;
- Labels on both the **x-axis** and **y-axis**;
- a Key, if you have used colours or symbols;
- been drawn neatly and accurately with a ruler.

**Figure 19** This graph contains some errors. Compare it to Figure 7 on page 44.

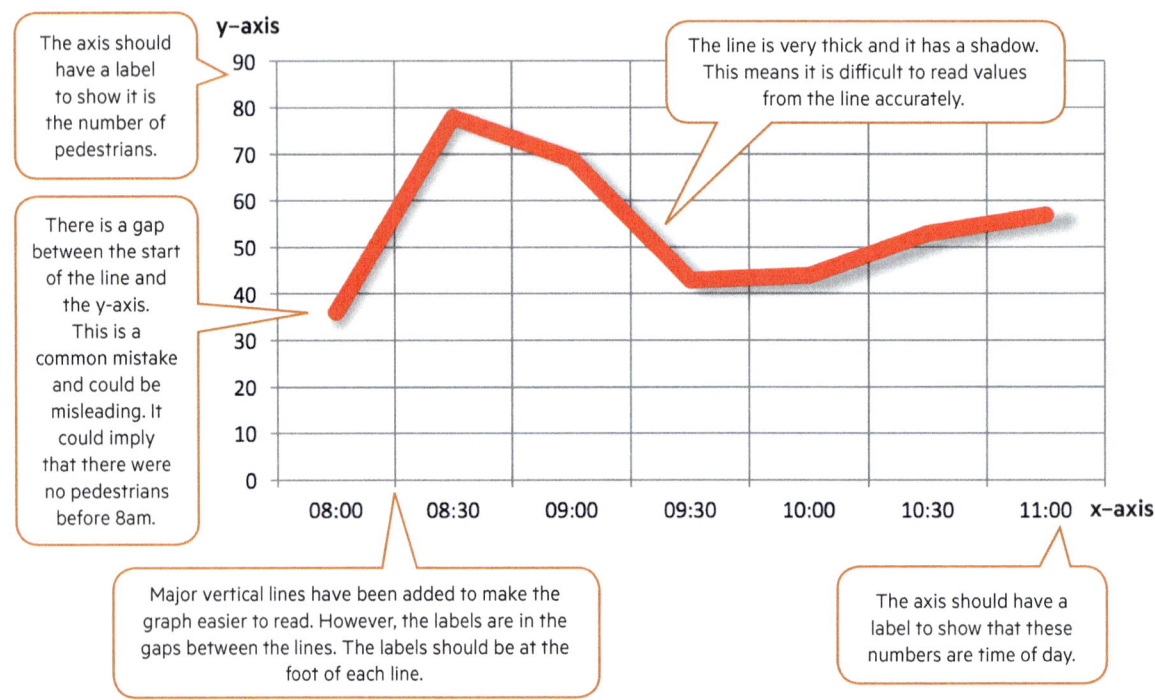

The axis should have a label to show it is the number of pedestrians.

There is a gap between the start of the line and the y-axis. This is a common mistake and could be misleading. It could imply that there were no pedestrians before 8am.

The line is very thick and it has a shadow. This means it is difficult to read values from the line accurately.

Major vertical lines have been added to make the graph easier to read. However, the labels are in the gaps between the lines. The labels should be at the foot of each line.

The axis should have a label to show that these numbers are time of day.

## Dos and don'ts of drawing graphs

**Do:**

✓ Label each axis and check that you have included the unit of measurement (for example, cm or m/s) in your label.

✓ Start the origin of the y-axis at zero unless you have a good reason not to.

**Don't:**

✗ Include a gap between the start of a line graph and the y-axis.

## Activities

1 **Explain why you should use a thin line to join data on a line graph.**

2 **Suggest why the y axis of a bar chart should usually begin at zero.**

3 **Study the list below. In each case, suggest an alternative style of chart or graph. Explain why your choice is suitable.**

   a) A line graph showing the number of pedestrians at 10 different locations around a town centre.

   b) A pictogram showing how wind speed changes with altitude.

   c) A histogram showing how people responded to a bipolar survey.

   d) A pie chart showing discharge measurements for 10 equally spaced sites along a river.

# Misleading graphs

A good graph allows you to see the data clearly. If the graph is badly drawn it could be misleading – making it difficult to see patterns or make comparisons. One common mistake is to provide a false origin for the y-axis – in other words, one that does not start at zero. Study Figure 20. This pair of graphs shows what happens if the y-axis does not start at zero. Students had collected data on pedestrian flows in a busy shopping street. The left hand graph gives the impression that more than twice as many pedestrians were seen at site B than at site C. However, the y-axis starts at 60 not zero, so differences in height between the bars has been exaggerated. In the right hand graph the y-axis starts at zero. You can see that site A has more pedestrians than B, and B has more than C. However, the differences are actually quite small.

 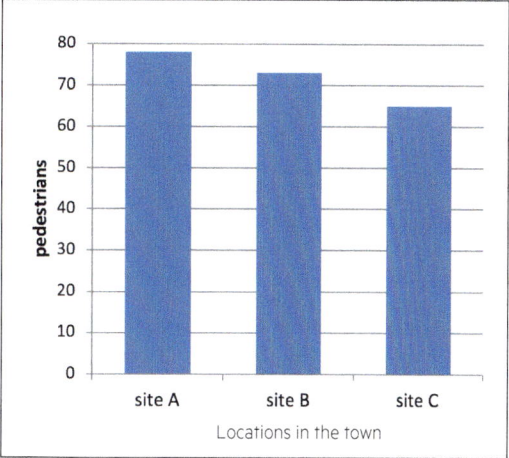

**Figure 20** Misleading bar charts.

# Pictograms

**Pictograms** are simple pictures that are used to represent quantities. The picture is drawn in proportion to the quantity it represents. Study Figure 21. On the left the pictogram is drawn correctly. Each pictogram represents 10 cars, so location A had 45 cars. On the right the pictogram is drawn incorrectly. The value is 3 times greater so the car has been enlarged until it is 3 times its original length. However, this is misleading. The car is also 3 times its original height. This means the pictogram is actually nine times larger than the original pictogram, suggesting that it represents 90 cars rather than 30.

**Figure 21** Misleading pictograms.

# Drawing maps

Geographers are interested in spatial patterns. These are patterns that exist in two dimensions so they can be represented on a map. Can you represent some of the evidence from your fieldwork on a map? You can use either **primary** or **secondary data** to draw your map.

## Desire line maps

A **desire line** map uses thin straight lines to show how places are linked together. In geography fieldwork they are often used to show the movement of people, for example, commuters travelling to and from work or visitors to a honeypot site. To draw desire lines you need to know exactly where each line begins so you need to ask the commuter or visitor the name of their home town rather than the name of a county.

**Step One** Download or sketch a map to use as a base map.

**Step Two** Plot a straight line for each item of data. In Figure 22, the line should begin where the visitor lives and end at the location of the honeypot site.

**Step Three** Give your map a **scale line** and north arrow. This is essential so you can describe the pattern made by the desire lines.

**Figure 22** A desire line map showing visitors to Dovedale. The area coloured green shows the Peak District National Park.

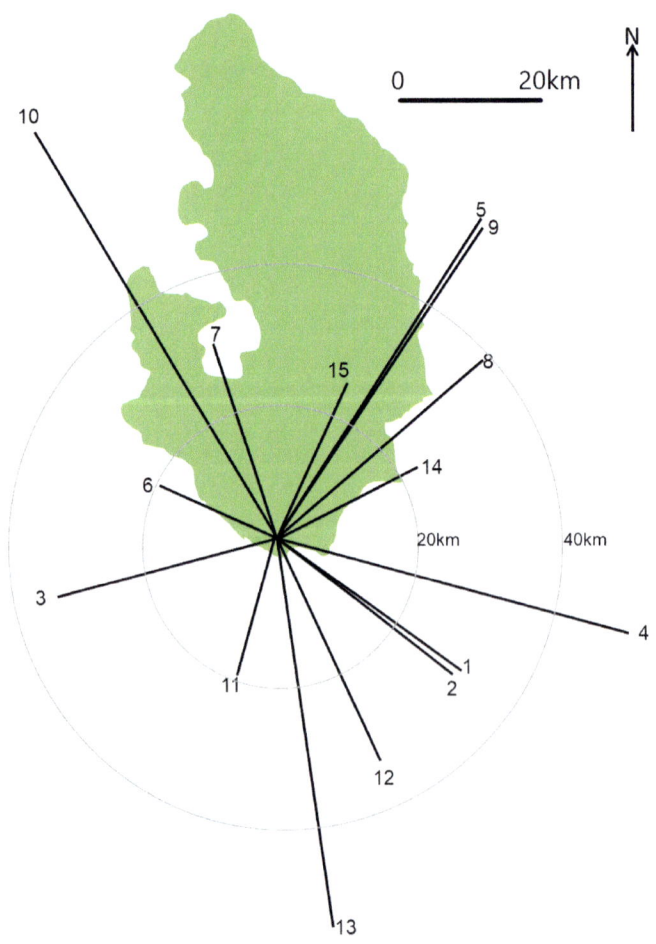

# Flow line maps

A **flow line map** uses proportional arrows to show flows, for example, flows of traffic or the movement of pedestrians. The width of each arrow is in proportion to the value of the flow. Proportional arrows can be used to show:

- flows from an area (such as a county) or a place (such as a town);
- data that has an amount and direction but where you don't know where the movement began (for example, traffic on a road).

**Step One** Draw a simple base map.

**Step Two** Create a scale for the width of your arrows. This needs to be small enough so that your largest arrows will fit on the map.

**Step Three** Draw each arrow onto the base map so that they point in the direction of flow. The width of the arrows should be in proportion to the data.

**Step Four** Add a scale line, north arrow, and key to your map.

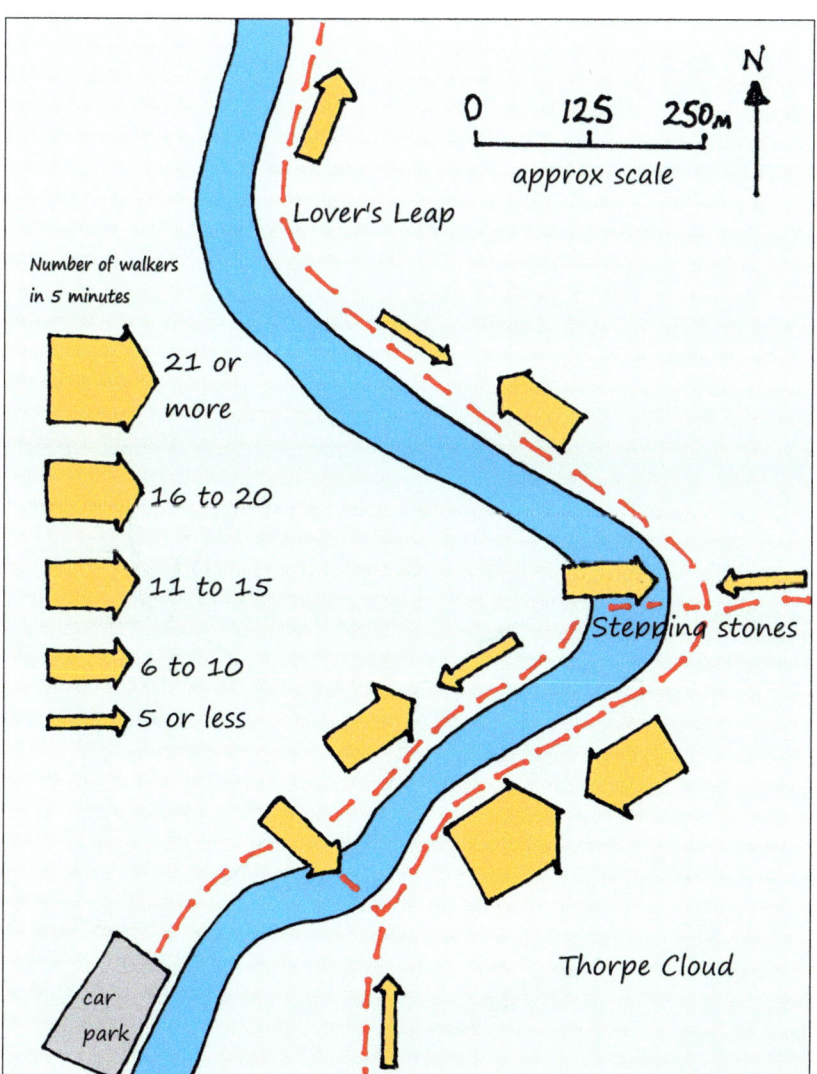

**Figure 23** Proportional arrows show the number of people walking on footpaths in Dovedale, Derbyshire.

- How to draw maps with proportional symbols.

# Proportional symbols

A **proportional symbol map** uses symbols (usually squares or circles) of different sizes to represent the data. The larger squares/circles represent larger data values – it is the area of the symbol that is in proportion to the value.

**Step One** Calculate the square root of each piece of data. Put the number in to a calculator and press this √ button. Add your answers to a table like Figure 24.

**Step Two** Decide on a scale for the symbols. The square root for Raven Meadows is 29.3. A simple scale would be 1:1 so, for Raven Meadows, you would draw a square with a side that is 29mm long, or a circle with a radius of 29mm. If this scale doesn't work well you could consider halving (to make all symbols smaller) or doubling (to make all the symbols larger) all of the square root numbers in your table.

**Step Three** Download or sketch a simple map to use as a base map.

**Step Four** Draw each symbol. The centre of each symbol should be located carefully in the right place. It is ok to overlap symbols.

**Step Five** Draw two symbols of different sizes at the edge of your map to make a scale. Use whole numbers that are close to the top and bottom of your range of data.

**Step Six** Give your map a scale line and north arrow.

**Figure 24** Number of parking spaces available in Shrewsbury (2018).

| Car park | Number of spaces | Square root √ |
|---|---|---|
| Bridge Street | 48 | 6.9 |
| Raven Meadows | 856 | 29.3 |
| Saint Julian's Friars | 272 | |
| Wyle Cop | 240 | |
| Frankwell | 720 | |
| Shrewsbury Station | 189 | |

**Figure 25** Proportional squares showing number of parking spaces available in Shrewsbury (2018).

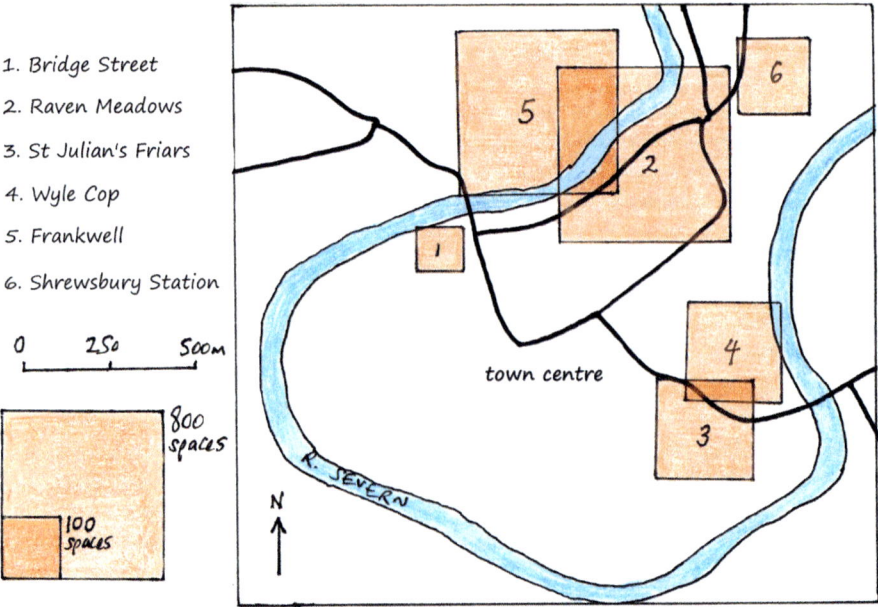

1. Bridge Street
2. Raven Meadows
3. St Julian's Friars
4. Wyle Cop
5. Frankwell
6. Shrewsbury Station

0    250    500m

800 spaces

100 spaces

town centre

R. SEVERN

N

# Located bar charts

Proportional symbols are quite tricky to draw. If your range of data is small, an alternative technique is to draw located bars on a map like Figure 26. However, located bars have a disadvantage - as data values increase, a bar chart gets taller and the largest data needs very tall bars that would be difficult to fit on the map. Proportional symbols also get bigger as data values increase but, because it is their area rather than their height that shows the increase, they remain compact enough to fit on the map. So proportional symbols are useful if you have a large range in your data.

https://en.parkopedia.co.uk/
Use this GIS to find the number of parking spaces near to your fieldwork site.

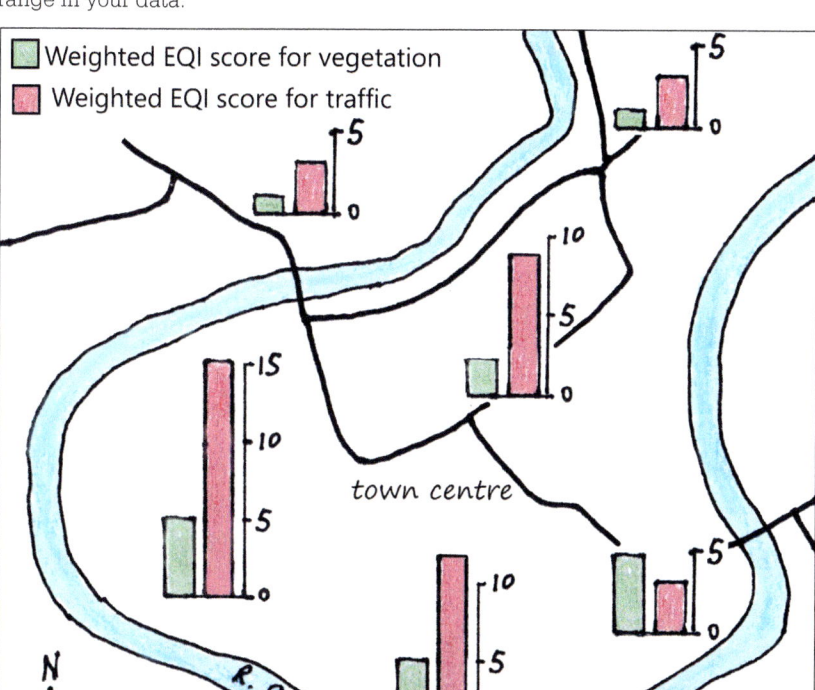

**Figure 26** Located bars on a map representing bipolar scores.

| Strengths and limitations of proportional symbols | |
|---|---|
| **Some strengths** | **Some limitations** |
| • Proportional symbols are useful if you have a large range in your data. | • It is difficult to estimate the size of a proportional symbol by eye – people underestimate their size. |

## Activities

1 Evaluate Figure 26. Identify at least one strength and one weakness.
2 Explain why the car park spaces data are best shown using proportional symbols rather than located bars. Use evidence from Figure 25 to support your answer.

# Chapter 4

# How can evidence be analysed?

### Learning objectives
- How to analyse trends on a graph.

https://www.riverlevels.uk/esk This website publishes discharge data for rivers in the UK. You can use it to see how the discharge of a river changes with time. It is too dangerous to collect river velocity data when the river is in flood, so this website is particularly useful for seeing how river discharge changes after heavy rainfall.

## What is analysis?

Analysis is like detective work. You have collected a lot of data. Now you need to sift through this data carefully, selecting the evidence that reveals connections, patterns, or trends. This process will allow you to make sense of the data.

Like a detective, you will need to examine a variety of evidence. You may have collected evidence in the form of tables of data, photographs, interviews, graphs, charts, or maps. The techniques that will help you make sense of this evidence are described in this chapter.

As you begin to analyse your data it might be helpful to ask yourself some questions. For example:

- *Do I have any evidence for …?*
- *Why do I think that…?*
- *How clear is the pattern or trend?*
- *How strong is the connection between this variable and that?*
- *What may be causing this variation in the data?*
- *What effect does this variable seem to have?*

### Three key points

A simple technique that you can use with almost any kind of evidence is to identify the three key points that the data is telling you. If you are making sense of a graph or a table of data you might look for:

- the overall trend – is it going up or down?
- the range – the difference between the highest and lowest values;
- any values that don't fit the overall trend. We call these anomalies.

Figure 1 shows an example of a hydrograph. This is a line graph that shows how the amount of water in a river changes over time. If you want to challenge yourself:

- try to identify five key points instead of three;
- use descriptive words that describe the patterns or trends you can see. Figure 2 gives examples of useful words to use.

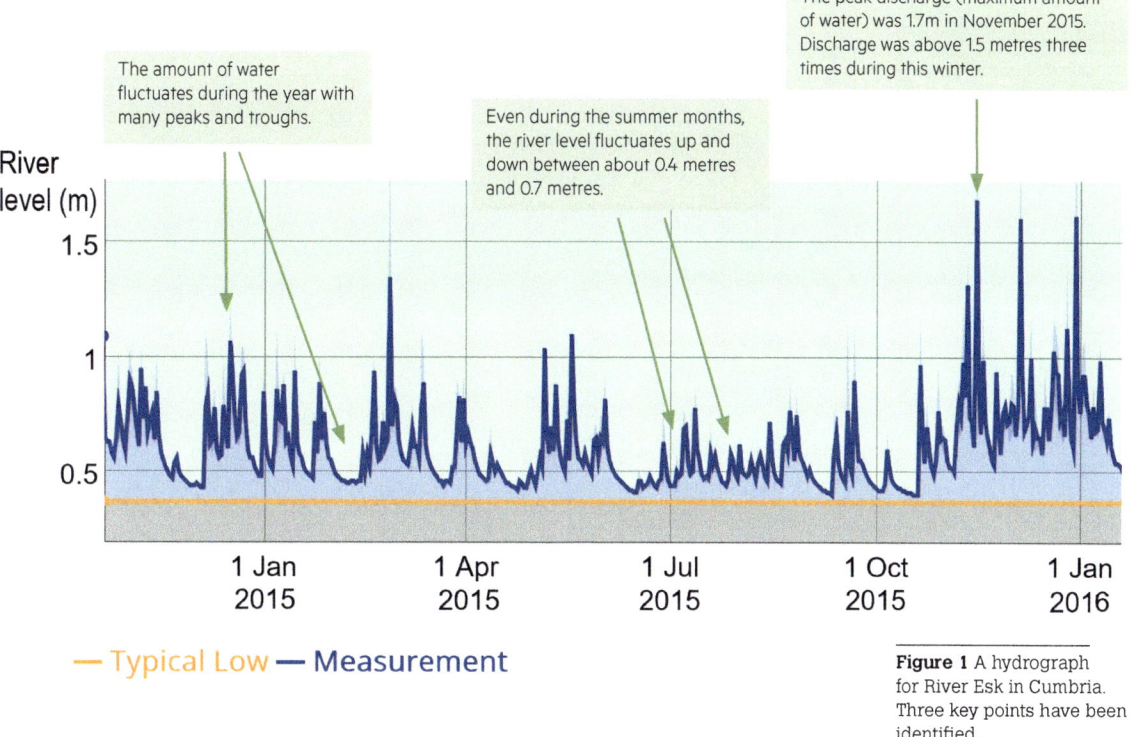

The amount of water fluctuates during the year with many peaks and troughs.

Even during the summer months, the river level fluctuates up and down between about 0.4 metres and 0.7 metres.

The peak discharge (maximum amount of water) was 1.7m in November 2015. Discharge was above 1.5 metres three times during this winter.

— Typical Low — Measurement

**Figure 1** A hydrograph for River Esk in Cumbria. Three key points have been identified.

| | | | |
|---|---|---|---|
| **Clustered** | Values (or points on a map) are concentrated into small groups. | **Linear** | Features (on a map or photograph) are spread out along lines. |
| **Decreasing** | The values are going down. Add **slowly**, **steadily** or **rapidly** to describe the trend. | **Random** | Values (or features on a map or photograph) are at irregular spaces and show no clear pattern |
| **Fluctuating** | The values in the data vary up and down in a repeated pattern. | **Regular** | Values (or features on a map or photograph) are repeated at even spaces. |
| **Increasing** | The values are going up. Add **slowly**, **steadily** or **rapidly** to describe the trend. | | |

**Figure 2** Useful descriptive words to make sense of data.

## Activities

**1  Add two more key points to Figure 1.**

# Text analysis

Text is an important source of evidence. It can provide both facts and opinions. The way that the text is written may tell you something about what somebody thinks of a geographical issue – their perception. You may have collected text as part of your evidence if you have interviewed people or asked open questions in a questionnaire. You may also have collected text evidence from books, newspapers, or the internet in the form of:

- blogs;
- interviews;
- adverts;
- news articles;
- reviews.

## Simple ways to analyse text

How do you analyse any patterns or trends when your evidence is in words? There are a number of simple things you can do.

1. Use one colour of highlighter to identify facts and another colour to identify opinions. In Figure 4, one fact has been highlighted in yellow in interview 1. One opinion is highlighted in green.

2. Consider the tone of the text – what kinds of words are used? Are the words mainly positive or is a negative tone used in some of the text? You can count the words or phrases that have a positive tone and then compare those words/phrases that have a negative tone in a bar chart. In Figure 4, one positive phrase has been highlighted in pink in interview 2. One negative phrase is highlighted in blue.

3. Copy and paste your text into an online tool that counts the frequency of words that are used. The text of the interviews in Figure 4 has been analysed in this way and the top 20 words used in each interview are shown in Figure 5. You can analyse these lists. Are the words mainly nouns that identify physical features of the environment, or human features? Are there many adjectives? The number of times that adjectives such as 'beautiful', 'pretty' and 'friendly' can be graphed using a bar chart. This will tell you something of the positive (or negative) perception of a place.

## Activities

1. **Analyse the text in Figure 4.**
   a) Identify one fact and one opinion in each interview.
   b) Count the number of times that each interview describes:
      i) a physical feature of the environment;
      ii) a human feature of the environment.
   c) What are the main similarities and differences between Interviews 1 and 2? Use evidence from Figures 4 and 5. You could draw bar charts to help this analysis.
2. **Describe how the interviews in Figure 4 could be used as a pilot survey to create a more closed qualitative survey. See page 23 to help your answer.**

**Figure 3** Borth, a small seaside town in west Wales.

**Interview 1** Borth is a small seaside town in West Wales. It is a wild and natural place that is close to nature. The estuary is constantly changing: the light shines on the water and reflects off pools of water on the beach. Birds soar in the sky above. This place is magical. It's as though the landscape is alive. The light is beautiful. I think it's because the weather changes so much and we have light that reflects off the estuary and the waves. The beach is glorious. When the tide is out a long way the wet sand glistens and shines in the light. The sand is soft and clean. I think the beach is really beautiful. I love walking along the pebbles looking for things that have been washed up in the tide. The beach is a lovely place but it can be wild in the winter when the storm waves crash against the pebbles. Then you know that nature can have wild forces.

**Interview 2** Borth is a friendly place. Local people are very friendly and want to help. I like the main street with its cheap cafes and shops selling holiday souvenirs. You can get most things in the local shops but not everything so we have to drive quite a long way to do some shopping - like clothes. The main street has pretty little cottages on either side. Some of the old cottages are made out of pebbles off the beach. The town has a lot of interesting history. There are special legends about the place too - legends about the sea and the drowned forest. I think the history of the place is very interesting. Borth is pretty in the summer and the beach is beautiful but it can be rather bleak here in the winter and sometimes I feel quite isolated.

| Interview 1 | | Interview 2 | |
|---|---|---|---|
| Primary words | Frequency | Primary words | Frequency |
| light | 4 | place | 3 |
| beach | 4 | legends | 2 |
| place | 3 | history | 2 |
| wild | 3 | beach | 2 |
| pebbles | 2 | cottages | 2 |
| sand | 2 | pretty | 2 |
| beautiful | 2 | shops | 2 |
| water | 2 | main | 2 |
| nature | 2 | Borth | 2 |
| estuary | 2 | friendly | 2 |
| shines | 2 | local | 2 |
| reflects | 2 | very | 2 |
| think | 2 | street | 2 |
| waves | 2 | quite | 2 |
| tide | 2 | interesting | 2 |
| glorious | 1 | isolated | 1 |
| much | 1 | sometimes | 1 |
| weather | 1 | old | 1 |
| landscape | 1 | beautiful | 1 |
| magical | 1 | cafes | 1 |

**Figure 4** Transcripts of two interviews. Students were investigating the concept of place in Borth. They interviewed visitors and recorded what they said about the place.

http://www.textfixer.com/
tools/online-word-counter.
php

**Figure 5** Text analysis of the two interviews to identify the most frequently used words

# Further ways to analyse text

If you can turn text into something else it can help you to analyse it and make sense of it. One simple way to represent text in another form is to count the key words and represent them in a bar chart – as suggested on page 58. Two other ways to represent text are shown in Figures 7 and 8. Both represent the text shown in Figure 6 below.

**Figure 6** Text from the Visit Wales website describing Borth. Source: http://www.visitwales.com

Borth's beach is three miles of gently shelving golden sand and is especially popular with families with younger children and sailboard enthusiasts. The tide goes out a long way, so its shallow waters are great for the little ones to paddle in and splash about.

Dog restriction on section of beach 1 May – 30 Sept.

At the southern end of this wonderful beach, an ancient submerged forest is exposed by the ebbing tide. Welsh legend has it that the trunks and tree stumps of old forests, long hidden under sand and sea, are the remains of the land of Cantre'r Gwaelod, which disappeared under the waves of Cardigan Bay, long ago.

Cantre'r Gwaelod was protected by dykes and dams but one night a feast was held and the place's night watchman called Seithennyn became inebriated and neglected his duty, resulting in the submerging of Cantre'r Gwaelod. After a storm and the ebbing of a very high tide, more and more of these ancient tree stumps come into view.

Blue Flag and Seaside Award beach. Toilets, cafes, restaurants, pubs, shops, parking. Lifeguard service provided 2 July to 4 September. Dog restrictions apply May to end September.

## Word clouds

One way of representing text in another form is to create a **word cloud** (or wordle). You could use this type of analysis to explore how the identity of a place is perceived. A word cloud (or wordle) gives a very visual way of recording how people see and identify a place. The size and position of each word in the 'cloud' is in proportion to its importance in the original text. For example, in Figure 7 (which has been created using the text in Figure 6), the word 'beach' is large and in the centre of the word cloud because it is the most commonly used word in the original text passage. This enables you to see what people think about the place and how those thoughts shape their perception of identity.

**Figure 7** A word cloud based on the text in Figure 6.

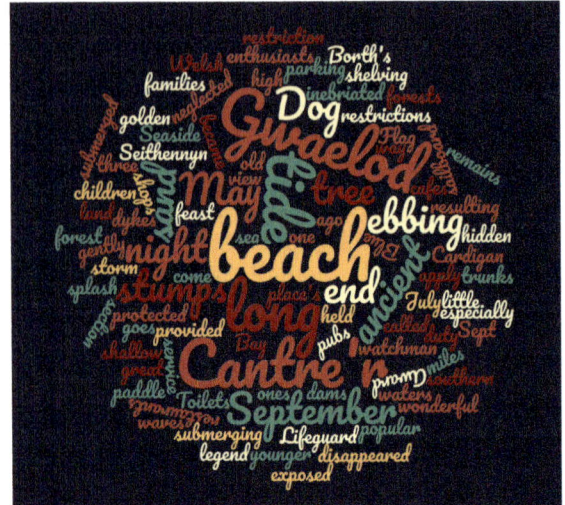

A word cloud can be made by following these steps.

**Step One** Decide on an interview question that matches the focus for your investigation, or select text from an internet site.

**Step Two** Interview your target group (this might be related to age / occupation / gender). The more people you ask the better.

**Step Three** Decide if the word cloud is designed to represent images of identity before a visit / after a visit (or both if the aim is to compare images and identities before and after visits).

**Step Four** Input the text from all of those interviewed into a suitable website to produce the word cloud. The website allows you to change font and colour variation to add further visual impact.

## Create spider diagrams

Word clouds are visual and simple to make. However, making a spider diagram to represent your text is a more sophisticated way to analyse the evidence. You can still pick out key features of the text by placing them in the centre of your diagram. The beach is a key feature of Figure 6 so it is central in Figure 8. The main advantage of using a spider diagram is that you can use it to identify the connections that bring meaning to the text. For example, Figure 8 has identified the physical and human features that make Borth special. These are picked out in the red clouds. It then uses connections to explain why these features are important. The most important conclusions of this analysis have been highlighted in orange.

Several websites allow you to create word clouds. Here are two:
http://www.wordle.net/
http://www.wordclouds.com/

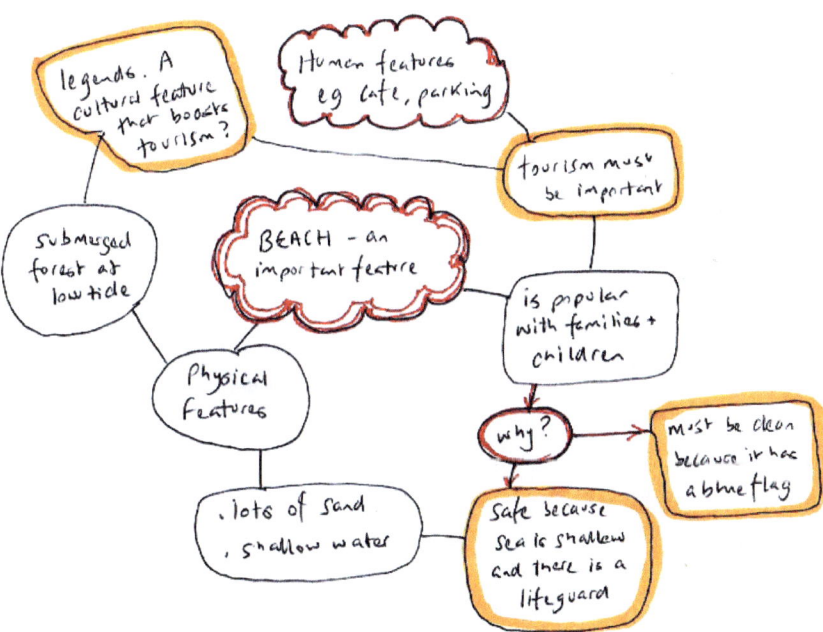

**Figure 8** A spider diagram based on the text in Figure 7.

## Activities

1  **Use a search engine to find a description of Borth. Choose a website that either encourages people to visit or reviews a visit to Borth.**
   a) Copy the text into an online text analysis tool and graph the results.
   b) Create a word cloud for your text.
   c) Analyse it using a spider diagram.
2  **Which of these analysis techniques seems to be the most useful? Justify your choice.**

# Making connections

To make sense of your data you also need to make connections between the evidence. Looking for evidence of cause and effect will help you to explain the evidence. For example, imagine an investigation of the shopping centre shown in Figure 9. You might investigate the patterns of pedestrian flows. Why are some parts of the town centre busier than others? You could collect evidence such as:

- pedestrian surveys at different times of day;
- **interviews** with shoppers about the features of the town centre that they like and dislike;
- the rateable value of shops could be found on the internet.

To make sense of this evidence you might look for connections such as:

- whether the busiest streets also have a lot of features that the shoppers like;
- whether rateable values are higher in the busiest streets. If so, this might explain why national chain stores are located on busy high streets but local shops are on quieter side streets.

**Figure 9** Why is this shopping street particularly busy? Shrewsbury.

**2** Do chain stores like these only locate on roads that have the greatest footfall?

**1** Are people attracted by the benches and flowers?

## Activities

1   a)   **Suggest three different reasons why some shopping streets have more pedestrians (footfall) than others.**

   b)   **Identify the data that you would need to collect to investigate the reasons for variations in footfall in a busy shopping street.**

# Scatter graphs

A **scatter graph** is a type of graph that is used to show the possible connection (or correlation) between two sets of data. To draw a scatter graph you will need **bivariate data**. That is two sets of data (or variables) that you think may be connected in some way. To draw a scatter graph follow these steps.

**Step One** Think about how the two variables may be connected. For example, in Figure 10, a student has collected data from the Census about the educational background and occupations of people living in 14 different neighbourhoods in Cardiff. The student thinks it is likely that more people will have professional occupations in **wards** where more people have a degree. If so, the percentage of people in professional occupations will be the **dependent variable** because the value of this number will depend on the percentage of people who have degrees – not the other way around.

**Step Two** Draw a pair of axes. The dependent variable should go on the vertical axis. The **independent variable** – in this case the percentage of people with a degree – should go on the horizontal axis. Try to make the two axes about the same length as each other so that the finished graph is square. This will make it easier to see any patterns on the graph.

**Step Three** Label each axis.

**Step Four** Plot the points for each ward onto the graph. Vertical crosses are the best way to plot each point. These will line up with the vertical and horizontal grid lines on your graph paper – so your plots should be accurate and easy to read.

**Step Five** Calculate the **mean (M)** value for each variable. Plot this point onto your graph – perhaps using a different style or colour so that it stands out from the other points on the graph.

**Step Six** Draw a line of best fit on your graph. The line must pass through the mean value (M). It should also follow the trend of the other points – but it doesn't need to go through any of them or join them up. However, there should be the same number of points on each side of the line of best fit (see page 65). Notice that the line of best fit does not have to go through the **origin** of the graph.

## How do I tell if there is a correlation?

A scatter graph will show a correlation between the two variables if:

- the dependent variable increases as the independent variable also increases. This pattern suggests that there is a **positive correlation** between the two variables. Figure 11 shows an example of a positive correlation;
- the dependent variable decreases as the independent variable increases. This pattern suggests that there is a **negative correlation** between the two variables. Figure 12 shows an example of a negative correlation.

There is no correlation between the two variables if the points are scattered across the graph without making any pattern.

| Ward | % of residents with a degree | % of residents in professional occupations |
|---|---|---|
| Adamsdown | 28 | 14.5 |
| Butetown | 33 | 20.1 |
| Caerau | 10 | 5.4 |
| Canton | 50 | 32.8 |
| Cathays | 22 | 13.4 |
| Ely | 9 | 6.1 |
| Grangetown | 22 | 13 |
| Llanishen | 31 | 19.5 |
| Pontprennau | 38 | 23.8 |
| Rhiwbina | 35 | 27.9 |
| Riverside | 32 | 20.1 |
| Rumney | 13 | 9.2 |
| Splott | 26 | 17 |
| Whitchurch | 37 | 26.8 |
| Mean | 27.57 | 17.83 |

**Figure 10** Census data for education and occupation in selected wards of Cardiff.

**Figure 11** Scatter graph representing the relationship between the bivariate data in Figure 7.

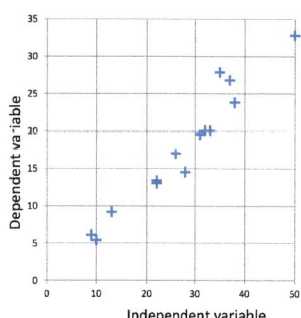

**Figure 12** Scatter graph showing a negative correlation.

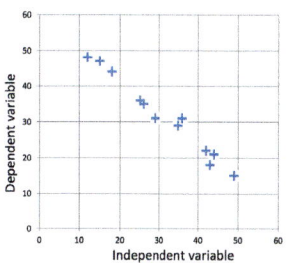

# Correlation and causality

**Correlation** implies that there is some connection between two sets of variables. However, the explanation for this connection may not be clear. For example, in studying the Census data for Cardiff, a student noticed that the percentage of households in inner urban wards who did not own a car was higher than in suburban wards. The student plotted a scatter graph with distance from the CBD on the horizontal axis. The graph showed a negative correlation. However, the explanation for this link is not clear. It could be that:

■ Households in the inner city are poorer and cannot afford a car.
■ Fewer homes in the inner city have driveways so there are fewer places to park a car.
■ People living in the inner city have access to frequent buses and trains so don't need a car.

**Causality** means that there is a direct link between the two variables. This link can be used to explain why the value of the dependent variable changes. For example, in a study of 20 European cities, a student found a negative correlation between average July temperatures and latitude, that is distance from the Equator.

## Activities

1 **Study the list of paired variables in the table below. For each pair state:**
   a) which is the dependent variable and which is the independent variable;
   b) whether you would expect to see a positive or negative correlation;
   c) whether you think the relationship shows correlation or causality. Make sure you can justify each of your decisions.

| Fieldwork context | Variable One | Variable Two |
|---|---|---|
| A sandy beach | Wind speed (metres per second) | The height that sand is carried (by the wind) above the beach (in centimetres) |
| A transect up a hillside between 150m and 400m above sea level | Height above sea level (metres) | Wind speed (metres per second) |
| A pebble beach | The average size of sediment (millimetres) | Distance along the beach – measuring in the same direction as the movement of longshore drift (metres) |
| An eroded footpath in a honeypot site | Distance from the car park (metres) | Width of the area affected by erosion (metres) |
| Study of 10 different soil samples | Infiltration rate (cubic cms per minute) | Percentage of the soil composition made of clay particles |
| The area around a busy main road | Noise levels (decibels) | Distance from the road (metres) |
| A housing area close to an attractive urban park | Distance from the park (metres) | House prices (£1,000s ) |
| A study of a river channel over several days | The discharge in the river channel (in cubic metres per second) | The amount of rainfall each day (in millimetres) |
| A larger urban area | Population density (number of people living per km²) | Distance from the city centre (km) |
| A larger urban area | Population density (number of people living per km²) | The percentage of households who do not own a car |

2 **Use Figure 13.**
   a) Interpolate a value for % of residents in professional occupations if 42% of residents have a degree.
   b) Using the other values on this graph, suggest how accurate your interpolated value is likely to be.

# Extrapolation and interpolation

You can use a line of best fit to estimate other values that could be expected if other points followed the same trend. This process is called:

- **Interpolation** if you estimate a value that is within the range of values plotted on your scatter graph. If most points on your graph are close to the line of best fit then interpolation should give a reliable estimate. However, the estimate will be less reliable if some points lie further away from the line of best fit.
- **Extrapolation** if you estimate a value that is outside the range of values plotted on your scatter graph. Extrapolation should give a reasonably accurate estimate if you try to estimate values that are only a little beyond the range of figures on your graph. However, you need to be cautious and consider whether your estimate is realistic. For example, on Figure 13, a ward with 2% of residents with a degree would have a minus figure for residents in professional occupations – this would be impossible!

The value 54 on the independent axis lies outside the range of plotted points. If the best fit line is extended (as it has been here) then a value on the dependent axis can be estimated at 39. This is an example of **extrapolation**.

The value 20 on the independent axis lies within the range of plotted points. A value on the dependent axis can be estimated at 11. This is an example of **interpolation**.

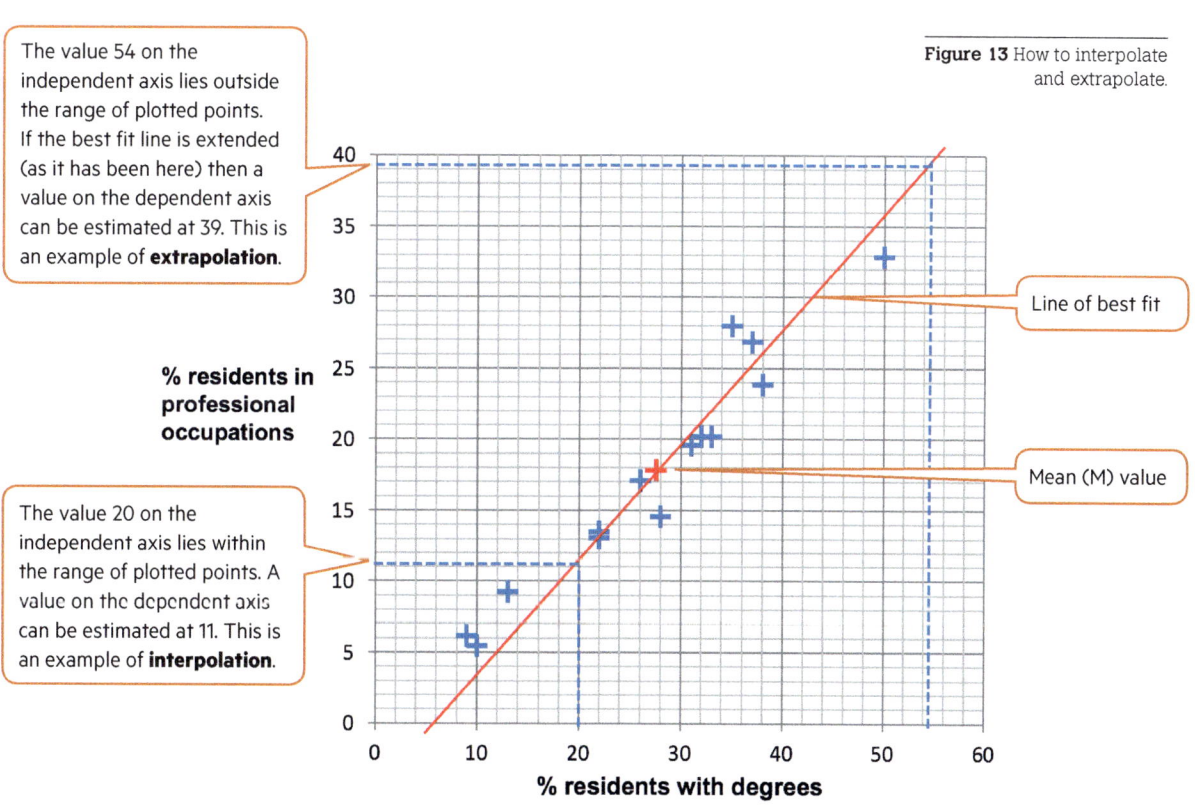

**Figure 13** How to interpolate and extrapolate.

% residents in professional occupations

% residents with degrees

Line of best fit

Mean (M) value

| Strengths and limitations of scatter graphs | |
|---|---|
| **Some strengths** | **Some limitations** |
| • Scatter graphs can give a visual clue about the correlation between two sets of data with different scatters for positive and negative correlations.<br>• A line of best fit can help you see the strength of the correlation. The closer the points are to the line, the stronger the correlation appears to be. | • If you only have a few points (less than 10) you shouldn't draw a scatter graph because what appears to be a pattern in the points could have been created by chance.<br>• Just because you can put a line of best fit through some scatter points doesn't mean you can assume that there is cause and effect (a causal link). |

# Chapter 5
# Conclusion

## Learning objectives

In this chapter we will explore:

■ How to reach conclusions from your fieldwork enquiry.

# How to reach a conclusion

You've collected data, processed, presented, and analysed it. It's time to end your enquiry by making your conclusions. A conclusion should make use of the evidence you have collected and analysed. To conclude an enquiry you need to:

1 remind yourself of the enquiry's aims;
2 consider the evidence;
3 draw together your overall findings for each enquiry question or hypothesis;
4 make a judgement/construct an answer to your original aim.

**Step One** Revisit the aims of the enquiry.

It is important to remind yourself of the aims of your enquiry. The aims of an enquiry are often based on your understanding of a geographical issue, concept, or theory. What was the concept that you were investigating? What did you expect to find – did you make any predictions? In your conclusion you need to decide whether or not you have met your aims.

**Step Two** Consider the evidence.

Think about each piece of evidence from primary and secondary sources. Each separate table, graph or map you have drawn is a piece of evidence and each tells a small part of the story. What patterns, trends, and connections can you see?

**Step Three** Draw together your overall findings.

You will need to draw together all of the separate pieces of evidence from your enquiry. It is rather like making a jigsaw puzzle. You need to bring these pieces of evidence together so that you provide an overall conclusion.

**Step Four** Write your conclusion.

When you write a conclusion you should refer to each piece of evidence. Remember to answer each hypothesis. Draw it all together by making a judgement, constructing an answer to your original aim.

## Messy geography

The results of your fieldwork will not always be what you expected. When you look at the evidence, graphs may show no real trend and patterns on maps are not always neat and tidy. For example, you might think that a map of house prices for your local town would show that the lowest house prices are close to the town centre. You might also expect that your fieldwork would show that house prices get steadily more expensive as you move out towards the suburbs. If so, a map of house prices would show a regular (or systematic) pattern. In reality, such a map is likely to be much more complex. It is likely to show clusters of higher house prices in inner urban areas – probably connected to positive features of the environment such as urban parks. In reality, the results of geography fieldwork are often messy rather than neat and tidy. Don't be surprised if your results don't exactly match your predictions.

**Figure 1** A photograph of the study area. Borth, West Wales.

A group of students investigated the effect of new coastal defences at Borth, a small seaside town in West Wales. They wanted to know:

- whether the coastal defences had positive or negative impacts on the town;
- how large the sphere of influence might be.

They conducted bipolar surveys to see what positive and negative impacts the construction of the new coastal defences might be. They also collected evidence of house prices to see whether the coastal defences created enough confidence for people to ask high prices for homes that were close to the sea. Their hypotheses were:

- *House prices are higher closest to the sea.*
- *Houses at sea level are more vulnerable to flooding (despite the defences) so are worth less than houses above sea level.*
- *The sphere of influence is affected by height above sea level rather than distance from the sea.*

Two key pieces of evidence are presented in Figures 2 and 3.

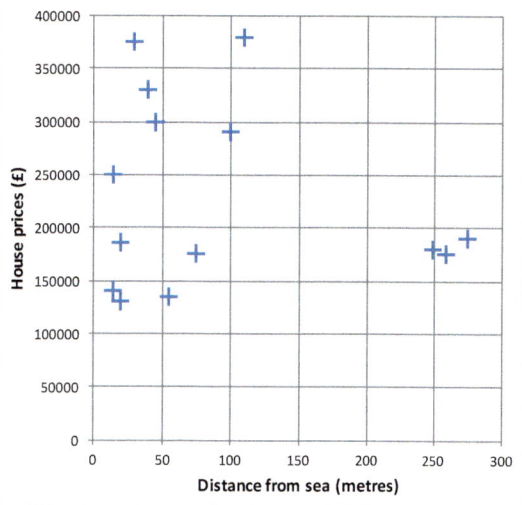

**Figure 2** A scatter graph showing the possible relationship between distance from the sea and house prices in Borth.

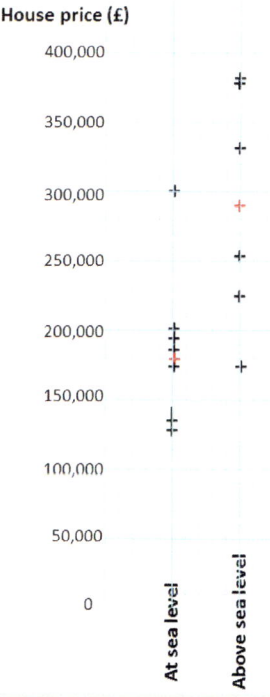

**Figure 3** A dispersion graph showing the range of house prices at sea level (less than 3 metres above high tide) and above sea level (on the hillside in Figure 1).

## Activities

1 **Describe the evidence shown in Figure 2.**
2 **Use Figure 3 to identify:**
   i) the median value of house prices in each location;
   ii) the range of house prices in each location.
3 **What conclusion can you draw when you put these two pieces of evidence together? Think about whether the students were able to answer any of their hypotheses.**

# Chapter 6
# Evaluation

# Identifying limitations in data

Evaluation is the final stage of the enquiry process. It is an opportunity to weigh up the strengths and weaknesses of what you have done. Your conclusions are based on the data so if there are problems with the data, or faults in the way that data has been collected, then the conclusions could be wrong. That's why it is so important to think critically about whether the data can be trusted.

There are two main types of limitation with data.

**Accuracy** - how close a measurement is to the actual value (or true value). For data to be accurate it needs the equipment to be used correctly and the result to be recorded carefully in the correct units.

**Reliability** - how certain you can be that your measurements are true. For data to be reliable it needs to be collected in such a way that, if the fieldwork is repeated, a similar set of data would be collected.

## Limitations in primary data

There are a number of reasons why there may be errors in **primary data**. Accuracy and reliability of data may be affected by:

■ errors when you read the apparatus, for example, you are weighing some sediment samples but forget to zero the scales;

■ carelessness when the results are recorded. Data errors can be caused by reversing numbers when you write them down. Untidy handwriting can also cause data errors;

■ a flaw in the sampling method, for example, you only take one noise reading and it happens to be at a very noisy moment. Wind speed and noise levels vary constantly. You will get a more reliable measure if you take several readings and then calculate the mean.

Figures 1, 2, and 3 give three examples of how data errors can occur and how they can be avoided.

If you are using a ruler to measure the length of a pebble you are relying on your eye to judge where the measurement starts and ends.

Using callipers means you can measure the true value exactly without any guesswork.

Measuring in millimetres is more accurate than measuring to the nearest centimetre.

**Figure 1** Use the correct equipment to improve accuracy of data.

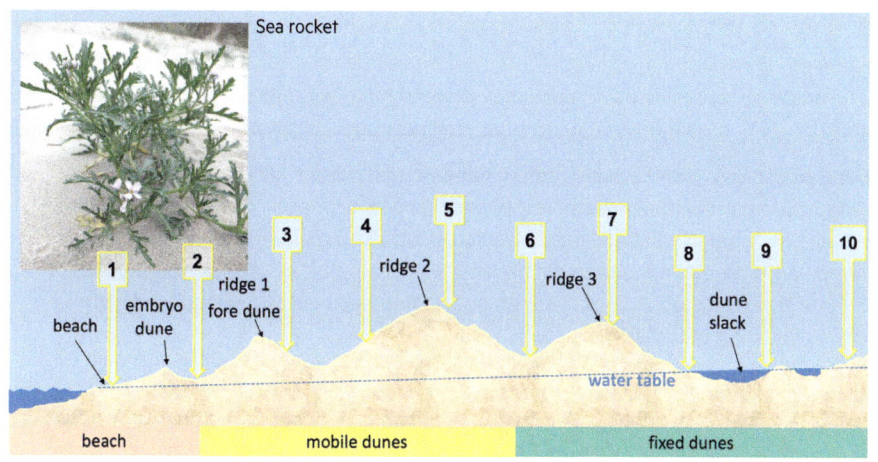

This systematic sample is a suitable technique. However, the transect is about 500m long so there are huge gaps between sample points.

Some variations in data occur over very short distances. For example, sea rocket only grows in the embryo dunes. If regular sample points are too far apart, this important plant will be missed completely.

**Figure 2** Use the correct sampling technique to improve accuracy of data.

Reliability is achieved by strictly following the same sampling procedures each time the data is collected. For example, unreliable data would be collected by teams of students if traffic was counted for 3 minutes at 10 locations but:

- the teams had not synchronised their watches so data was not collected at the same time;
- or some teams used stopwatches to keep to exactly 3 minutes but others did not.

**Figure 3** Use exactly the same sampling technique as other team members to improve reliability of data.

## Dos and don'ts of data recording

**Do:**

- ✓ Be neat when you write down results in the field. A lot of errors occur because numbers have been written badly and cannot be read later when you get back to the classroom.
- ✓ Check equipment has been zeroed correctly.
- ✓ Take multiple measurements if data is varying.
- ✓ Use exactly the same methods as other team members if you are collecting data in a group.

**Don't:**

- ✗ Wait until later (for example, after it has stopped raining) to record the data. You will forget the details so do it straight away.

# Limitations in secondary data

The accuracy of **secondary data** may depend on how recently the data was collected. For example, data for the national census is only collected every 10 years.

Some sources and websites are more reliable than others. We can trust that big data collected by a UK government department, or academic organisation, will be reliable because it will have used a suitable sampling method. However, data on some websites may be unreliable because we do not know who collected the data or how large the sample was. There may be **bias** because the website is trying to put across a certain point of view.

## Bias

Biased data may be factually incorrect, or its accuracy may be a matter of opinion. This is especially true of **qualitative** rather than **quantitative data**. If we analyse the evidence presented to us in the media (on TV, in newspapers, or in blogs) or in qualitative surveys such as **questionnaires** and **interviews**, we can identify that some data is presented as fact but is actually biased because it is someone's opinion. Identifying this bias is useful. It helps us describe the opinions of certain groups of people. The fact that qualitative data is sometimes biased is what makes it useful to us as geographers. This is because we may begin to understand why certain groups of people have strong opinions by considering what makes the evidence biased.

It is important to avoid writing questions that lead people to agree with your own point of view when designing a questionnaire. When you evaluate fieldwork that contains qualitative data collection you need to consider whether your method of data collection led to biased results or not.

## Suggesting improvements

As part of your evaluation you should think carefully about whether there is anything else you could have done. A lot of students write that they could have collected more data but think about what kind of data you needed. More of the same data might not have told you anything else. It might be better to think about other kinds of data and the issues you might have had collecting it. For example, could you have:
- asked different questions in a questionnaire?
- collected data at a different time of day?
- collected data at the same time but on another day?
- found some secondary data to clarify an issue where your primary data was inconclusive?

### Dos and Don'ts of evaluation

**Do:**
- ✓ Think about the strengths and weaknesses of your sampling strategy. With hindsight, did you choose the best strategy to give you data that represents the whole?
- ✓ Identify any possible inaccuracy or unreliability. Suggest how you might change the enquiry if you did it again.
- ✓ Identify bias and try to explain why certain groups of people provide biased evidence.

**Don't:**
- ✗ Simply suggest that you could have done more. A bigger sample might not have given you different results but different questions in your questionnaire might have been more effective.
- ✗ Write about what you enjoyed and didn't enjoy on the fieldtrip. That's not evaluation.

# Be SMART

You can evaluate your enquiry using the mnemonic SMART (Specific, Measurable, Achievable, Realistic, Timely). This technique, shown in Figure 4, can help you evaluate your aims and the design of your sampling techniques.

|   |   | 5 | 4 | 3 | 2 | 1 |   |
|---|---|---|---|---|---|---|---|
| S | The enquiry had a clear aim and specific focus. I know what I was trying to prove. |   |   |   |   |   | The aim was vague. It lacked focus. I didn't know what I was trying to prove or disprove. |
| M | I could measure all outcomes of the fieldwork. I used a range of techniques to collect qualitative and quantitative data. |   |   |   |   |   | I struggled to measure any outcomes of the fieldwork. I wasn't sure which techniques to use. |
| A | The enquiry was achievable and safe. Primary and secondary data was available to provide the evidence I needed to meet my aims. |   |   |   |   |   | There wasn't enough of the right kind of data available to me so I couldn't find the evidence I needed to prove / disprove my aims. |
| R | The aims were realistic. I had the right data collection skills and accurate equipment. |   |   |   |   |   | The aims were unrealistic because I didn't have the right skills or equipment to collect the data I needed. |
| T | I had enough time to collect data across a range of places / time so that I could see patterns and trends. |   |   |   |   |   | I would have needed more time to collect enough data to see any patterns or trends in the data. |

**Figure 4** Use this SMART bipolar survey to evaluate your enquiry.

These students measured river depth so that a cross section of the channel could be drawn. They measured down from a line that was stretched horizontally across the river channel.

They took depth readings at 1 metre intervals as they crossed the river. They used a rule to take the measurements and recorded the data to the nearest centimetre.

They didn't have any way of checking whether their line was actually horizontal. There was some slack in the line – you can see it in the photo. So, if they had repeated the task they would probably have found some significant variation in their measurements.

**Figure 5** How accurate and reliable are the results from this data collection?

## Activities

1 **Make a copy of Figure 4.**
   a) Use the bipolar statements to evaluate your own enquiry.
   b) Once you have completed the survey, you could:
      i) explain why one aspect of your enquiry was successful;
      ii) explain why another aspect of your enquiry had limitations;
      iii) suggest and justify how your enquiry could have been improved.
2 **Study the information in Figure 5.**
   a) Describe one strength of this fieldwork.
   b) Suggest one way this fieldwork could be made more accurate and one way it could be made more reliable.

# Chapter 7

# Investigating changing places

## Learning objectives

- Understanding the concept of place and how it might be investigated through fieldwork.

## What is the concept of place about?

Each place has features that give it a unique character. People recognise this character – it gives the place, and the people who live there, a sense of identity.

People often form strong attachments to a place. They identify with places where they were born, where they grew up, or where they live today. They may have all sorts of reasons for forming this attachment. They may identify with:

- **the people.** Perhaps a famous person was born here. Perhaps residents have a distinctive accent, or use dialect words and phrases that give the place a unique character;
- **the culture.** Perhaps the place is linked to different parts of the world through business and migration. Multicultural communities help create strong identities within UK places – see Figure 1;
- **the environment.** Perhaps the place has some special landmarks that people feel a connection with. These may be historic or modern buildings (like the Rotunda in Figure 2);
- **the economy.** Some places have strong connections with a particular industry such as motor cars, mining, steel making, or pottery.

**Figure 1** Caribbean market in Electric Avenue, Brixton.

## How can fieldwork help us investigate places?

Before you begin, you need to decide on an aim for your investigation of place. You might investigate one of these aspects of a place.

- How different groups of local people feel about the features that make a place unique, for example:

*Do residents of different ages have different points of view about their home town?*

- How and why a place changes over time.

*Do local people think these changes have been successful?*

- How the identity of a place is influenced by globalisation, for example:

*How important are global links created by trade and migration in helping to form the identity of a port?*

We can investigate the concept of place by:

- asking as many people as possible to list 3 things that they like best about a place. Use the words to make a **word cloud** (see page 60);
- asking 10 people to take 10 photos each of the place you are investigating. Sort the photos using categories such as historic features, buildings, parks and green spaces, street scenes, and people to see which features are photographed most often.

## Comparing media perceptions to local perceptions

It's easy to find out how a place is represented in the media. Do an online search for *'What is (place name) famous for'*? If you do this you will quickly come up with a list of buildings, landmarks, famous people, or local industries that are connected to your town. You may also find cultural references – search for Birmingham and you will probably find references to Balti (curry), Richard Hammond, and Ozzy Osbourne as well as jewellery and the canal system. You can investigate whether the image of your place in the media is similar to or different from the **perception** of local people by following these steps.

**Figure 2** Birmingham's Rotunda tower block is a famous landmark in the city.

**Step One** Do your online search. Try to find between 10 and 20 landmarks or features that are connected to your place. Print images of each one.

**Step Two** Show your images to a group of local people. Ask them to sort the images in order of importance to them. Then select the 5 images that best represent how the group perceives the place.

**Step Three** Ask the group to identify any important features of the place that did not appear in your online search. Ask them why these are important. Try to photograph these features and annotate them using phrases used by your participants.

You could extend this investigation further by conducting the **interviews** with two separate groups of people. For example, you could interview one group of elderly local residents and a second group of much younger people. How do the perceptions of each group compare with the media representation of their place? How do their views compare with each other?

## Activities

1. a) Do an online search to find between 10 and 20 landmarks or features that are famous or important in your own town.
   b) Look at the images of these features. How have they been photographed? Do they show a positive image? Is the sky blue? Are people smiling? Discuss why positive images are important for places.
   c) Discuss the list of features with your class. What photos would you take that would represent your local place?

# Using photographs to analyse change over time

One simple way to see how much a place has changed over a longer period of time is through photography. Our towns and cities change a lot when redevelopment takes place. Old industrial districts change as factories and warehouses are demolished and new buildings - often shops, flats, and offices - go up in their place. High streets change when shops change hands, or when traffic is banned to make more space for pedestrians. These changes usually take place over several years so you will need to compare today's urban environment with secondary data to be able to analyse the changes.

A great source of secondary data is old photographs, especially old postcards. These are available, to view or buy, on online auction sites. Once you have found some old photographs of your town you need to find the spot where they were taken. Then re-photograph the same place as it looks today. This technique is sometimes called **re-photography**.

Figures 3 and 4 show the same street in Cardiff. Today, this street is one of Cardiff's main pedestrianised shopping streets. Notice how much the street has changed for pedestrians. You can also see that several buildings have been replaced. We can analyse this type of evidence by annotating the photographs. **Annotations** are more complex statements than simple labels. A good annotation would compare the two photographs. It might go on to suggest why the changes have happened. For example, in Figure 4, on the left of the street you can see a Santander Bank. This building, which is 7 Queen Street, can also be seen in Figure 3. You can see that the first and second floors have hardly changed but the shop front at street level is completely different. You could add an annotation like this:

> The shop front has a modern design and the brand name is clearly displayed. This is important because Santander is part of a large chain so they use their name and logo to promote an image of a modern bank to their customers.

## Activities

1 Study the features that are numbered 1-5 on Figure 4. Write annotations for each of these features.
2 Use both photos to describe how Queen Street has changed to make it a more attractive environment for shoppers.

7 Queen street

QUEEN STREET, from WEST CARDIFF

**Figure 3** Queen Street in Cardiff in 1946.

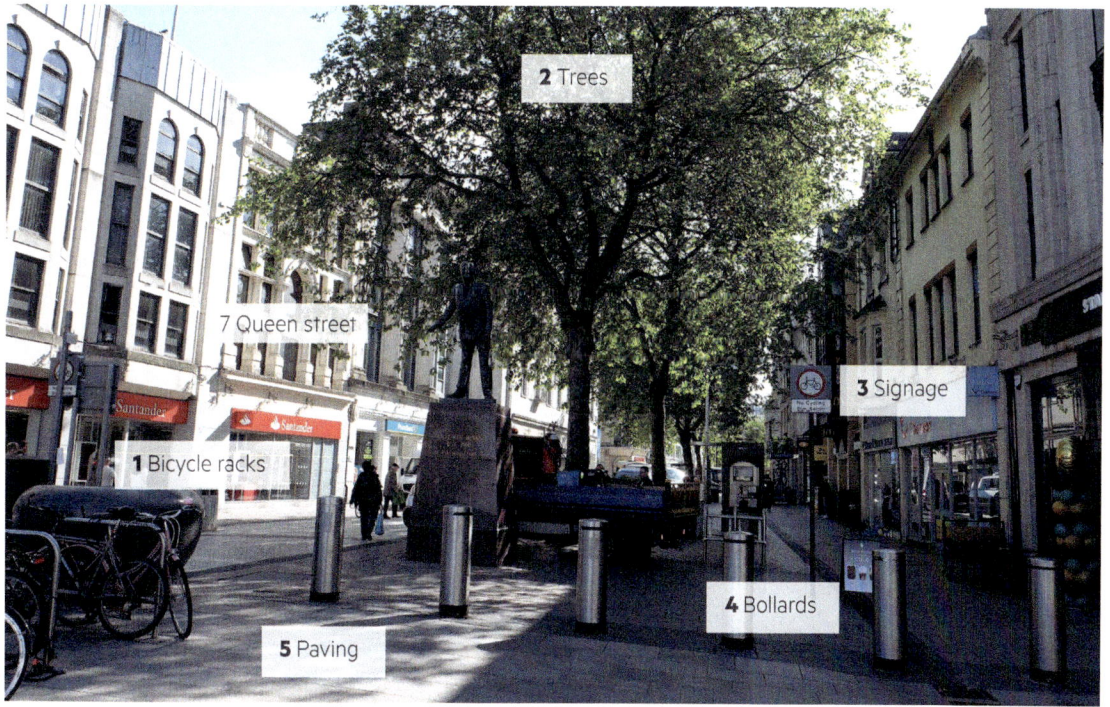

**2** Trees

7 Queen street

**3** Signage

**1** Bicycle racks

**4** Bollards

**5** Paving

**Figure 4** Queen Street in 2017.

# Investigating urban regeneration

The UK's towns and cities are constantly changing in a process known as urban regeneration. As older buildings become derelict they are demolished creating brownfield sites that are ready for redevelopment. This is an opportunity for planners to create an attractive urban environment and encourage investment by businesses. Carefully planned regeneration may solve problems such as dereliction and traffic congestion. Ugly buildings can be replaced with cutting-edge modern designs. Traffic systems can be redesigned to make streets safer for pedestrians and cyclists.

## How can fieldwork help us investigate regeneration?

A fieldwork **investigation** of urban regeneration needs to start with a clear **aim**. For example, you might decide that you want to investigate:

■ whether local people feel that the regeneration has been successful. If so, you will need to design some **questionnaires** or **Likert surveys**. You might also be able to use secondary evidence to find out people's opinions about their town using online blogs;

■ whether regeneration has had a positive effect on the urban environment. If so, you could **survey** areas of the town that have been redeveloped and compare them to areas that have not.

**Figure 5** The entrance to the Bullring Shopping Centre – an area that has been regenerated.

We can use fieldwork to investigate how regeneration has affected the urban landscape by:

■ measuring **footfall** (the movement of pedestrians) to see whether areas that have been regenerated have more pedestrians than other parts of the town/city;

■ using **EQI surveys** to assess whether regeneration has created a higher quality urban environment than parts of the town/city outside of the regeneration zone;

■ using **bipolar** or **Likert Surveys** to assess what people think of the urban environment.

## Activities

1 Use Figure 7 to describe the main similarities and differences between locations A and B. Remember that the axis for this graph includes negative scores.

# Using radial graphs to analyse urban environments

**Radial graphs** are a useful way to represent the data from an EQI or bipolar survey. This type of graph has several axes – in fact, you can have as many as you like – one for each variable. Figure 7 shows an example. It has been drawn using the results of a bipolar survey. This survey used 8 bipolar statements to compare two locations in Birmingham so the graph has 8 axes. Location A is an area that has been recently regenerated whereas location B is outside the regeneration zone. We can see that the area covered by the polygon for location A is considerably larger than the polygon for location B, meaning that the overall impression is that the urban environment is much more favourable at A than at B.

**Figure 6** The area around Moor Street station is just outside the regeneration zone.

To draw a radial graph like Figure 7 you can use a software programme like Excel. Radial graphs can also be drawn by hand. To draw a radial graph by hand follow these steps.

**Step One** Work out how many axes you need and the angle between them. To do this, divide 360 by the number of axes. For example, to draw Figure 7, 360/8=45, so the angle between each **axis** is 45 degrees.

**Step Two** Decide on a scale for the length of each axis. For example, starting at -3 (the lowest number) in the centre of the diagram, you could mark each axis at 1cm intervals. Do this carefully so that each axis is exactly the same length.

**Step Three** Plot the scores for each bipolar survey on each axis. If you are comparing two or more locations, use two or more colours.

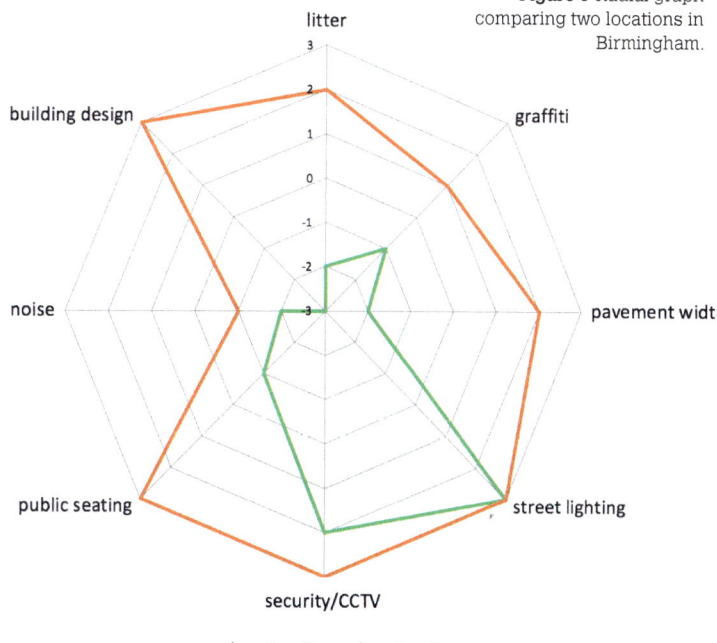

**Figure 7** Radial graph comparing two locations in Birmingham.

| Strengths and limitations of radial graphs | |
|---|---|
| **Some strengths** | **Some limitations** |
| • Radial graphs are very useful for making comparisons between two or more places.<br>• They allow you to represent more than one variable at a time.<br>• They are useful for plotting bipolar scores. | • Radial graphs have a limited use. They only really make sense if the different variables can all be measured using the same scale. |

# Investigating footfall

The UK's high streets are changing rapidly. The growth of out-of-town retail parks and online shopping have had massive impacts on high street retailers. Regeneration can make town centres more attractive so that more people, of different ages, continue to use our high streets. Regeneration schemes use a number of ways to do this, for example, by:

- making streets safer for pedestrians with wider pavements, traffic calming, and CCTV;
- increasing the attractiveness of the urban environment by creating small green spaces, improving signage, or adding public seating;
- adding more things for people to do besides shopping, such as places to eat and meet each other, as well as services such as libraries.

**Footfall** is a measure of how many pedestrians are visiting a place. It is usually simple to count people walking past over a period of, for example, 5 minutes. If you divide into teams of three and spread evenly through the town you can count footfall at different locations at the same time. This will give you the data you need to draw a map, like Figure 9. It is important that each group counts for the same length of time and at the same time. Otherwise your results will not be **reliable**.

**Measuring** footfall in a public open space, like a park or pedestrian area, can be tricky because of the number of people and the complexity of their movement. For example, they may change direction, or walk in groups with few gaps between them. One strategy to improve the **accuracy** of the count is for three people to count at the same location for five minutes. Then, calculate the **average** (see pages 40-41).

**Figure 8** It can be difficult to count footfall accurately in open spaces.

# How to draw an isoline map

An **isoline** is a line on a map that joins places that have an equal value. In fieldwork about changing places you could create an isoline map to show:

- **patterns** of footfall or litter across the town centre;
- noise levels around a building site, stadium, or busy road.

To draw an isoline map you will need lots of points across a map where data has been collected. Once you've got enough data you can draw the map – a bit like a dot to dot map. However, you also need to be able to **interpolate** - which means to find values that lie between other values.

Study Figures 10 and 11. Each dot on Figure 10 represents a place where footfall data has been collected. To keep it simple, we have rounded these numbers up or down to the nearest 10.

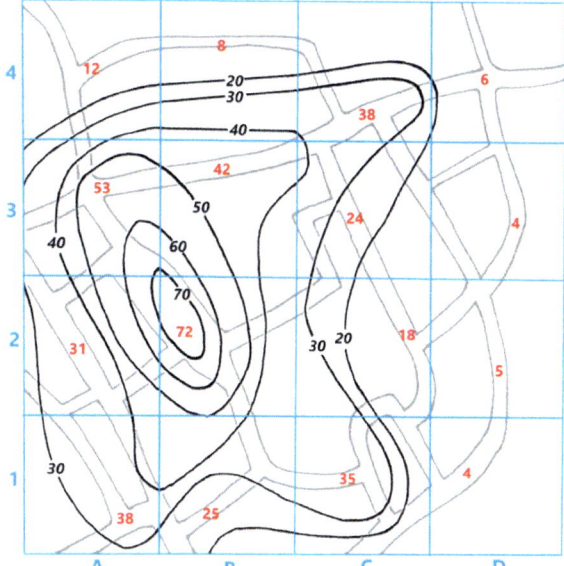

**Figure 9** An isoline map of footfall. Each red number represents footfall counted at that location. To collect data for this map, the students divided into 12 teams – one for each grid square of the map.

Study Figure 10. There are 5 places where a value of 20 was recorded. We could join them up but it would be easier if there were more points. To create more points we need to interpolate. Each of the red dotted lines has been drawn between values where 20 should be somewhere on the line. Using this method gives us another 5 places that have a value of 20. Figure 11 shows how to interpolate the data and draw an isoline for 20 pedestrians. Everywhere inside this line has a higher footfall.

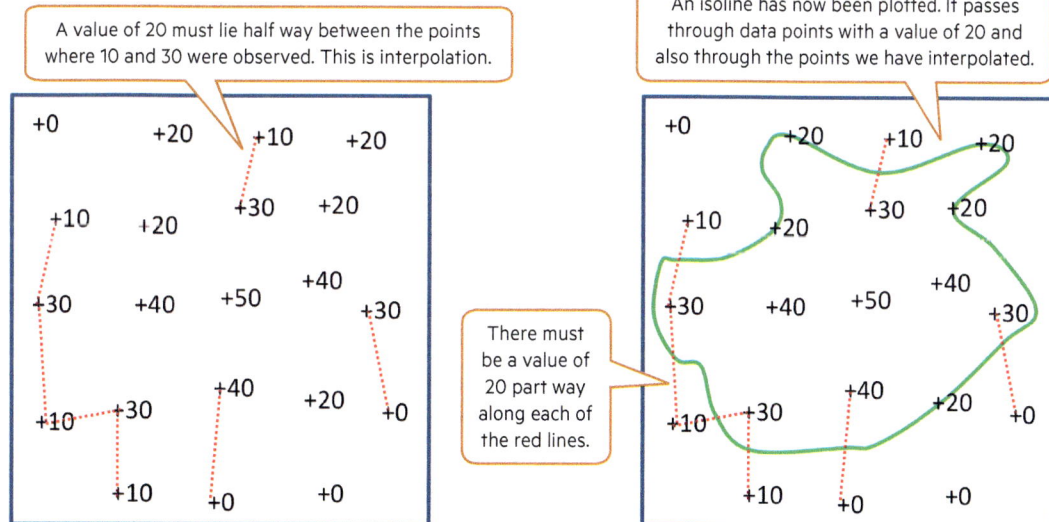

> A value of 20 must lie half way between the points where 10 and 30 were observed. This is interpolation.

> An isoline has now been plotted. It passes through data points with a value of 20 and also through the points we have interpolated.

> There must be a value of 20 part way along each of the red lines.

**Figure 10** Interpolation of extra data points with a footfall value of 20.

**Figure 11** An isoline with a footfall value of 20 has now been plotted.

## Activities

1. Explain why you need to coordinate your teams carefully if you want to collect footfall data across several locations in a town centre.
2. Make a copy of the data points in Figure 10. Use the interpolation technique to draw an isoline for the footfall value of 40.

# Investigating opinions about regeneration

Regeneration can change people's **perceptions** of an older urban area – making it seem more modern, friendly, and business-like. However, opinions are sometimes divided. When the Elephant and Castle Shopping Centre (seen in Figure 12) opened in 1965, the concrete and glass architecture was criticised by many. In 2018 it was announced that the controversial building would be demolished. Many local people protested. They feared that the regeneration would not include enough affordable housing. It seems that regeneration can be controversial – something that could be investigated.

**Figure 12** The Elephant and Castle Shopping Centre.

**Figure 13** The Selfridges building in Birmingham's Bullring Shopping Centre.

In Birmingham, the Rotunda tower (seen in Figure 2 on page 73) was opened in 1965. The adjacent Bullring shopping centre was redeveloped in 2003 with the inclusion of the Selfridge's building. The modern design, seen in Figure 13, has won eight awards. We can use fieldwork to investigate whether different groups of people have differing opinions about regeneration. To do this, follow these steps.

**Step One** Decide on your **enquiry question**. For example, do younger or older people prefer modern architecture?

**Step Two** Design a **questionnaire**, **bipolar survey**, or **Likert Survey**.

**Step Three** Design a data collection sheet. The **tally** chart, Figure 15, has been designed to be used with the bipolar survey, Figure 14.

**Step Four** Identify the **mode** in the results. In Figure 15, the mode for the question about architecture is +3.

**Step Five** You could present the data in a located **bar chart**, like Figure 16.

**Figure 15** A tally chart for a bipolar count.

| Tally for Centenary Square | | |
|---|---|---|
| Bipolar score | architecture | pedestrians |
| +3 | ⱵⱵ ||| | / |
| +2 | ⱵⱵ | |
| +1 | || | ||| |
| 0 | || | ⱵⱵ |
| -1 | ||| | // |
| -2 | | ⱵⱵ || |
| -3 | | || |

| Positive | +3 | +2 | +1 | 0 | -1 | -2 | -3 | Negative |
|---|---|---|---|---|---|---|---|---|
| Interesting architecture | | | | | | | | Boring architecture |
| Safe and easy for pedestrians | | | | | | | | Dangerous and difficult for pedestrians |

**Figure 14** A student's bipolar survey.

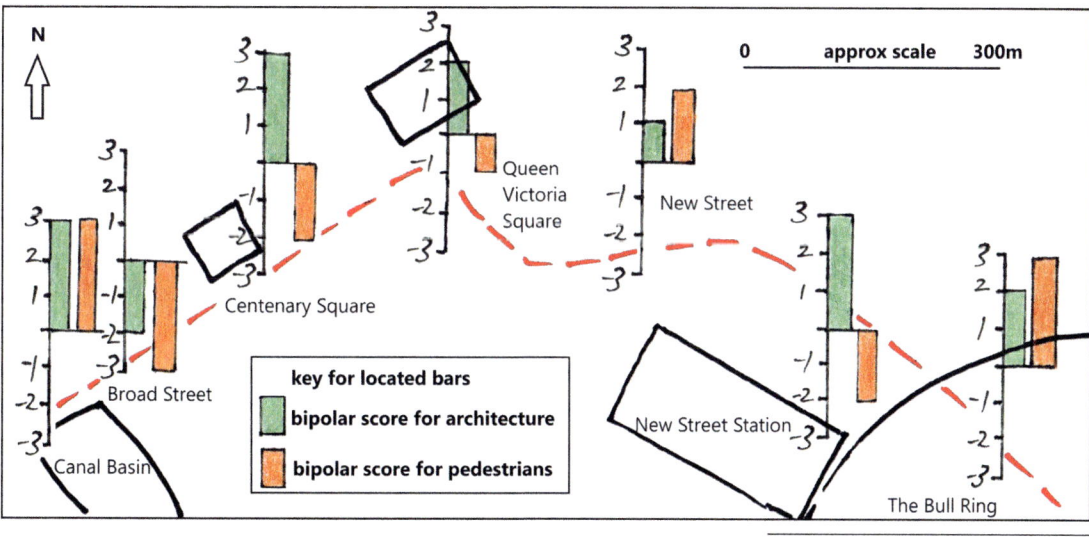

**Figure 16** A student's located bar map of a transect through Birmingham.

## Activities

1 **A student investigated regeneration on a transect through Birmingham.**
   a) Identify **one** weakness in the bipolar survey (Figure 14). Suggest how it could be improved.
   b) Suggest another pair of bipolar statements that could be added to Figure 14.
   c) Discuss Figure 16. How useful is this map in helping us understand opinions about regeneration?

# Time to reflect

**Figure 1** The library in Centenary Square, Birmingham. Opened in 2013, this was a flagship building in the regeneration of Centenary Square.

**Figure 2** A student's questionnaire.

1 Tick the boxes that best describe your opinions about the regeneration of Centenary Square and Paradise Circus.

| | Strongly agree | Agree | Neutral | Disagree | Strongly disagree |
|---|---|---|---|---|---|
| The new library is more attractive than the 1970s building it replaced. | | | | | |
| Centenary Square is a nice place to spend some time. | | | | | |

2 In your opinion, is Centenary Square a safe place for pedestrians and is the signposting to other parts of the city clear?

3 What do you think are the main benefits of urban regeneration in Birmingham?

**Figure 3** The redevelopment of Paradise Circus and the east end of Centenary Square in 2018.

1 **Study Figure 2.**
   a) Identify two weaknesses in the design of this questionnaire.
   b) Explain why using this questionnaire could lead to unreliable results.
   c) Suggest one way that the questionnaire could be adapted to make it more reliable.

2 **During the redevelopment of Centenary Square in 2018, part of Broad Street (the street through the square) was closed to traffic. Pedestrians and cyclists were able to use a narrow route through the square that can be seen in Figure 3.**
   **A student investigated the impact of the closure of the street during the redevelopment. The student set the following hypothesis:**

   *'Pedestrians are generally in favour of the road closure but cyclists think it is a nuisance.'*

He used a Likert Survey to test the hypothesis. The details of his sampling strategy are shown in Figure 4.

| | Total number seen in the square in one hour | Number of people in the sample |
|---|---|---|
| Cyclists | 100 | 10 |
| Pedestrians | 2,000 | 50 |

Figure 4 Sampling strategy.

a) Identify one limitation in this sampling strategy.

b) If the student used a stratified sampling technique and asked 100 pedestrians, how many cyclists should he ask? Show your workings.

c) Give one reason why a stratified sample would have given more reliable results than that used by the student.

3 The results of the Likert Survey are shown in Figure 5. The student presented these results in the graph shown in Figure 6.

| 'The closure of the street to traffic was a good thing.' | Number of responses from... | |
| | Cyclists | Pedestrians |
|---|---|---|
| Strongly agree | 0 | 10 |
| Agree | 1 | 30 |
| Neutral | 1 | 7 |
| Disagree | 5 | 2 |
| Strongly disagree | 3 | 1 |

Figure 5 The student's results.

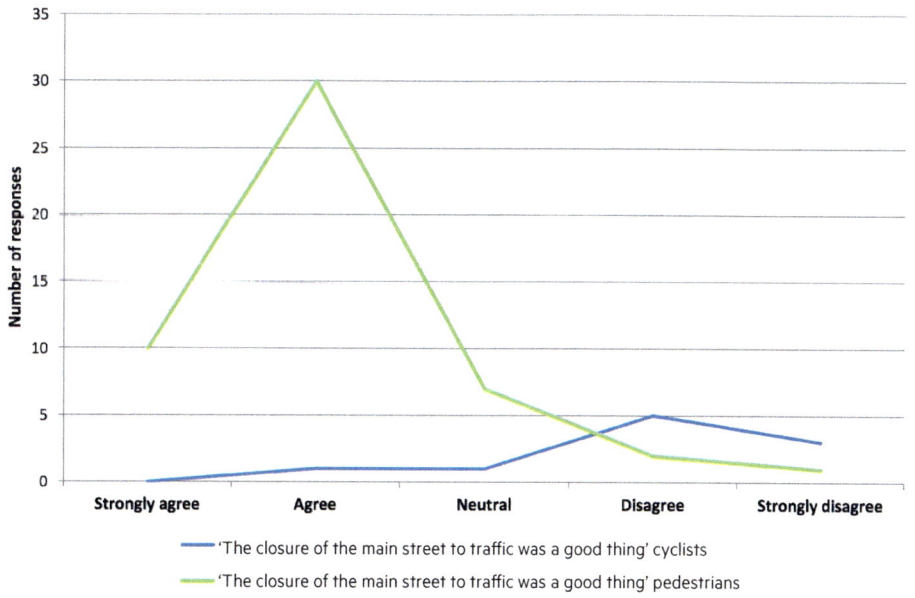

Figure 6 The student's graph.

'The closure of the main street to traffic was a good thing' cyclists

'The closure of the main street to traffic was a good thing' pedestrians

a) Use the evidence in Figure 5 to reach a conclusion about whether 'The closure of the street to traffic was a good thing'.

b) Evaluate the way the results are presented in Figure 6.

4 Reflect on your own fieldwork in a human geography environment.

Consider your planning. Were your aims SMART? Was the location suitable? What other enquiry questions could you have asked?

# Chapter 8
# Investigating sustainable places

## Learning objectives

- How to collect qualitative data about sustainable urban living.

## Investigating sustainable urban living

Poor housing, a lack of green space, traffic congestion, pollution, and crime are problems in some UK towns and cities. To make urban areas better places for residents (now and in the future) planners need to make the urban environment more sustainable.

Sustainable urban living means that people have access to local services that they need and that they have a reasonable quality of life that will last into the future. Residents of a sustainable neighbourhood are well-connected by a variety of transport links including buses, safe cycle routes, and pedestrian routes.

Sustainability is a big concept. Egan's Wheel (shown in Figure 1) breaks the concept of sustainability down into smaller, more manageable chunks. By using two or three of the features shown in Figure 3 we can create aims for fieldwork that are SMART because they are specific and measurable (see page 9).

**Figure 1** Egan's Wheel describes eight key features of sustainable living.

We can investigate sustainable urban living by assessing some of the key features shown in Figure 1.

- **Surveying** the neighbourhood - plotting the location of security features on a base map. You could plot the location of CCTV cameras, neighbourhood watch signs, and anti-climb paint.
- Using an **Environmental Quality Index (EQI) Survey** to assess the quality of green spaces, transport links, and access to services.
- **Interviewing** local residents or business people about features that should strengthen community spirit. How do people perceive the value of community centres, youth clubs, or support groups?

**Figure 2** This pedestrian area of Derby is designed with good lighting and strong materials that cannot be easily vandalised. CCTV cameras help to reduce crime and make local residents feel safer.

## Using Environmental Quality Indices (EQI) to assess sustainability

We can investigate sustainability using **EQIs** (see pages 28-29) to assess key features of the urban environment.

**Step One** Select two features from Egan's wheel and design an EQI like Figure 3.

**Step Two** Do a **pilot survey** using your EQI to make sure that it is possible for different students to apply the criteria in a consistent and reliable way. If not, re-write the criteria.

**Step Three** Design a **sampling strategy** that will allow you to spot any spatial variations in sustainability across the town/city. You will probably want to use a **base map** with grid lines (see pages 16-17). This is **systematic sampling**.

**Step Four** Make sure that everyone agrees on how to apply your EQI criteria so that the results are reliable. Decide who is going to sample each point and then collect your data.

**Figure 3** You can use an EQI to assess the sustainability of a neighbourhood.

| Feature | Criteria | | | | |
|---|---|---|---|---|---|
| | **5** | **4** | **3** | **2** | **1** |
| **Sensitive to the environment** | A wide range of different habitats are available to the public. | Private gardens and communal areas have trees/shrubs as well as grass. | Private gardens have trees/shrubs but communal areas are grass. | Some areas of grass, e.g. lawns, verges, or sports fields. | No green spaces. |
| | **5** | **4** | **3** | **2** | **1** |
| **Well-served** | A wide range of groups are actively involved in providing local community projects. | A wide range of services are available for different community groups. | Services are available for some different groups in the community. | Some services for local residents. | Very few services for any local residents. |

## Activities

1 Study Figures 1 and 2. Describe **three** features of the environment shown in Figure 2 that help to make this place sustainable.

2 a) Design an EQI to assess whether transport links are sustainable.

 b) Working in pairs, discuss the criteria you have used in your EQI. How might you adapt your EQI so that it could be used reliably by different students and still get a consistent result?

**Learning objectives**

■ How to use primary and secondary data to investigate impacts on well-being.

# Investigating impacts on well-being

Well-being is a measure of how happy and healthy we feel. Happiness can depend on a whole range of factors. Many of these factors are personal to us, for example, our physical and mental health, our home life, and our work life. We are also influenced by our environment. For example, it is well known that people generally feel happier when they are outside, especially if they are in the countryside or at the coast. This means that, if you live in a town, being close to green spaces, like Figure 4, can improve our well-being. People can be fearful of crime and their physical health can be affected by poor air quality. This means that features of the urban environment such as derelict buildings, vandalism, and traffic congestion can all have negative impacts on our well-being. There is a direct link between improving the sustainability of urban environments and improving well-being – a link that can be investigated through fieldwork.

**Figure 4** A public park in the centre of Cardiff, Wales.

We could investigate the impact of the park on local people by:

■ asking people what they think of the park using a **Likert Survey**. Use strong statements like *'It would be a great benefit to live close to this park'*. Survey people to see if they agree or disagree (see pages 26-27);

■ using information from estate agents to find local house prices. If the park really is attractive to people, then we might expect that houses overlooking the park are more expensive than similar houses a few streets away.

## Primary data collection

We can collect **primary data** to investigate aspects of the environment and whether they have good or bad impacts on our well-being. Compare Figures 4 and 5. People find green spaces attractive. They provide a safe place for exercise and relaxation so have a positive impact on the lives of local people. Now think about the impacts of the busy road. It may have some positive impacts for local people. For example, it means locals are well-connected to other places, especially if it is a bus route. However, traffic noise and pollution will have negative impacts for some local people. You could investigate a **hypothesis** such as:

*'House prices are higher in neighbourhoods that have more green spaces'.*

Figure 5 Traffic on a busy road in Birmingham.

We could investigate the impacts of the road by:

■ taking noise readings – perhaps using an app on your smart phone. How quickly do noise levels decrease as you move away from the road? Do noise levels vary throughout the day?

■ measuring the amount of dust in the atmosphere by sticking double-sided tape to buildings to collect dirt and dust on the open sticky side. Is there more dust close to the road?

■ collecting secondary data on house prices. Do busy roads have a negative impact on house prices?

## Using secondary evidence

People don't like giving personal information during a **questionnaire**, such as the value of their home. You will, therefore, need to use an online estate agent to find local house prices. Using online data like this has advantages, for example, you can very quickly collect data from a wide area which means you can see **spatial patterns**. However, be aware of any specific limitations too. House prices depend on a number of **variables**. These include location, type (for example, terraced or semi-detached), age, and size (for example, number of bedrooms). In this type of enquiry our main interest is in one of these variables, the location, because we want to know whether the features of the local environment have a significant impact on house prices. To make the results of the enquiry more **accurate** it is a good idea to compare the prices of similar sized houses in different locations, for example, only **sampling** the prices of semi-detached three-bedroom houses. This means that the variables of size and type are controlled, while we focus on the importance of location.

## Activities

**1 Imagine you wanted to investigate the effect of busy roads on house prices.**

　a) Choose a suitable enquiry question or hypothesis.

　b) Plan a suitable sampling strategy for this enquiry. Justify your plan.

# Investigating green spaces in the urban environment

The UK's towns and cities provide many opportunities for recreation and leisure. Green spaces provide places for relaxation, sport, dog walking, and exercise, for example, jogging. Sometimes recreation is organised, such as a park run.

Green spaces also provide safe routes for pedestrians and cyclists where they are separated from vehicles. Footpaths and cycle paths that follow canals or rivers, like the one in Figure 6, allow commuters to travel through the city quickly and safely. By creating safe pedestrian and cycling routes, planners can help to reduce traffic congestion. These routes can become part of an integrated transport system if they are linked to bus routes or train stations.

## Environmental benefits

Green spaces provide habitat for urban wildlife such as butterflies, moths, bats, garden birds, and mammals such as foxes. Trees help to reduce noise and traffic pollution. **Vegetation** helps to reduce the risk of flooding by intercepting water and reducing the time it takes for heavy rainfall to reach urban rivers.

Planners now encourage the creation of Pocket Parks – tiny green spaces that benefit the local community and the environment.

**Figure 6** Bute Park, Cardiff.

We could investigate the use of this green space by:

- observing how people use the space: how do people in different groups (such as teenagers, adults, families) use the space? Do these different groups use the space for different amounts of time?
- photographing any evidence that the park is being managed such as signs about cycling, ball games, or litter;
- using a **conflict matrix** to see whether any uses of the open space are incompatible with other uses;
- **interviewing** people who use the park (or Pocket Park) about the health or well-being benefits;
- searching social media to find people's views about the use of the park, such as attendance at a park run event.

## Using a conflict matrix

A conflict matrix, like Figure 8, is a simple way to record potential conflicts between different recreational uses of an urban green space. For example, in some parks, dogs must not be walked in areas designed for children's play. Some activities, such as cycling, could be perceived as a hazard, or even a nuisance, by other users of the park.

**Step One** Identify the main users and create a conflict matrix like Figure 8.

**Step Two** Collect evidence of any conflict. You could use Likert Surveys to identify the conflicts.

**Step Three** Record your findings in the conflict matrix using the traffic light key.

**Figure 7** A sign at the entrance to one of Cardiff's parks is evidence that different uses of the park can conflict with one another.

| | Walking | Jogging | Toddlers and children's play | Ball games | Dog walking | Cycling |
|---|---|---|---|---|---|---|
| **Walking** | | | | | | |
| **Jogging** | | | | | | |
| **Toddlers and children's play** | | | | | | |
| **Ball games** | | | | | | |
| **Dog walking** | | | | | | |
| **Cycling** | | | | | | |

**Figure 8** A simple conflict matrix.

Key

🟥 Conflict needs management

🟧 Some potential conflict between uses

🟩 Activities are compatible – no conflict

## Activities

1 **Think about a park close to your home or school.**
   a) Design a sampling strategy that could be used to count pedestrians in the park. Decide how many people you would need and where they might stand.
   b) Imagine your aim was to see whether family groups stayed in the park for longer periods of time than single people. Design a data collection sheet. It must allow the user to record evidence that will help answer this aim.

# Investigating high street sustainability

Many high street shops are chain stores - part of a national or multinational network of stores under the same ownership. Other shops, known as independent retailers, are owned by local business people. Many of these stock local produce, like the butcher shown in Figure 9. Mr Pugh buys all of his meat from farms less than 10km from his shop and it is slaughtered locally too. Some consumers prefer to buy this kind of produce. They want to support local businesses and they want their shopping to have a lower impact on the environment because of the lower food miles. This is a measure of the distance your weekly shopping has travelled before it arrives on your plate. Some would argue that buying local produce is an important way to make town centres more sustainable.

**Figure 9** Mr Pugh. An independent butcher in Bishop's Castle, Shropshire.

We can use fieldwork to investigate the global economic links of the high street by:
- plotting retail land uses and identifying multinationals, chain stores, and locally owned independent shops;
- measuring footfall (pedestrian flows) in different town centre locations to see whether there is any **connection** between **footfall** and rateable value.

## Investigating footfall and shop type

Retailers want their shops to be in locations that have high footfall where there are large numbers of pedestrians and therefore potential customers. These locations are highly desirable. Consequently, shop owners tend to pay a higher rate of tax (known as the rateable value) for these shops. This relationship may explain why many small, locally owned businesses can no longer afford to locate in the busiest streets of our town centres. Charity shops pay lower business rates so can operate in areas which independent shops can no longer afford.

**Figure 10** Costa is owned by Coca Cola, a US multinational company. It is the second largest coffee chain in the world after Starbucks – which is also a US multinational company.

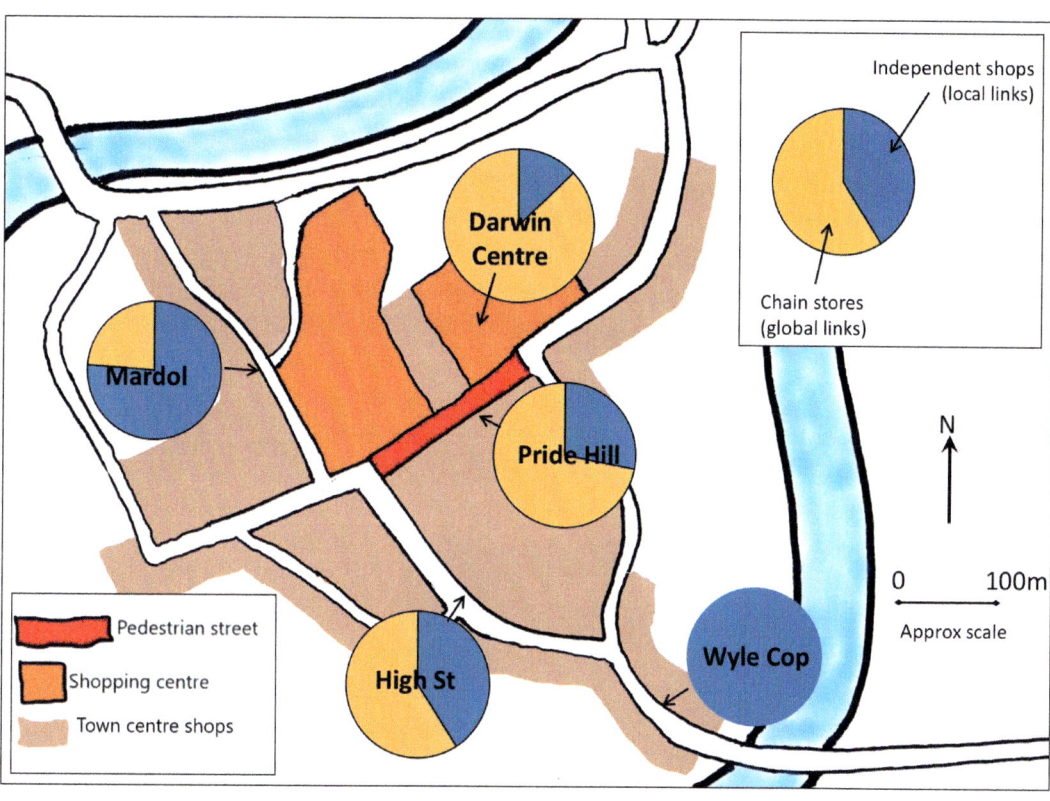

Figure 11 A student's map of Shrewsbury.

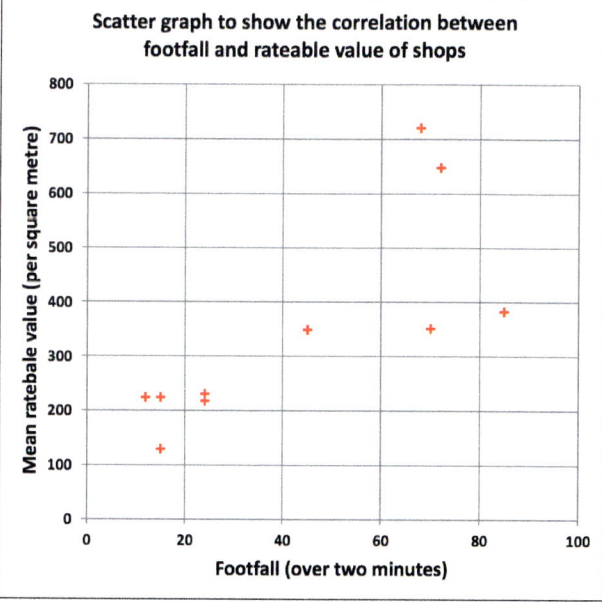

Figure 12 A student's scatter graph representing data collected in Shrewsbury.

https://www.gov.uk/correct-your-business-rates
This online tool allows you to find the rateable value of shops in your high street.

## Activities

1  **Study Figures 11 and 12.**
   a)  Describe the correlation shown in Figure 12.
   b)  To what extent does the evidence support the conclusion that independent shops are more likely to locate in pedestrian streets where footfall is highest? Use evidence from Figures 11 and 12.

- How to investigate issues of poverty and deprivation affecting rural areas.

# Investigating rural deprivation

If you live in the countryside it can be a long journey to the nearest shop, school, post office, theatre, or sports centre. Just getting to the nearest shop can be tricky because there aren't many buses. Consequently, many rural areas face issues that make them unsustainable places to live in. Some rural areas face challenging socio-economic issues too such as:

- an ageing population with a shortage of facilities for elderly residents;
- a lack of well-paid, full-time work.

**Figure 13** The closure of rural shops and services is a big issue in many rural communities. Bishop's Castle, Shropshire.

We can investigate issues of rural sustainability by:

- asking local residents about which services they use and how far they have to travel to use them;
- using **Likert surveys** to assess how locals feel about the closure of rural services;
- using **secondary data** from the census or a **GIS** site to collect evidence of rural poverty and deprivation.

## Coastal rural communities

**Figure 14** Choropleth showing multiple deprivation. Areas coloured blue/green have an above average standard of living. Areas coloured orange/red have a below average standard of living – these are areas suffering deprivation.

Some smaller seaside towns have particular problems with low incomes. Traditional fishing jobs have been lost. Jobs in tourism have declined as holidays abroad have become cheaper. Study Figure 14. Notice the **cluster** of orange and red areas close to the seaside town of Skegness and in rural Lincolnshire which indicate a standard of living that is below UK **average**.

# Secondary evidence of rural deprivation

It can be difficult to collect primary data about inequality or deprivation. Most people don't like talking about personal details so you should avoid questions about jobs, income, and health. Instead, we can use secondary data sources such as the **census**, or **GIS** maps, to collect social and economic data about the rural area. We can find data on house prices, crime, job types, qualifications, and the age structure of the rural population. Examples can be seen in Figures 14 and 15.

GIS maps can be used to help design data **sampling**. For example, we can use maps like Figure 16 to plan a route for a **transect** through a rural area. You could use **stratified sampling** – visiting one place of each colour on the map – and carrying out **EQI** surveys at each sample site.

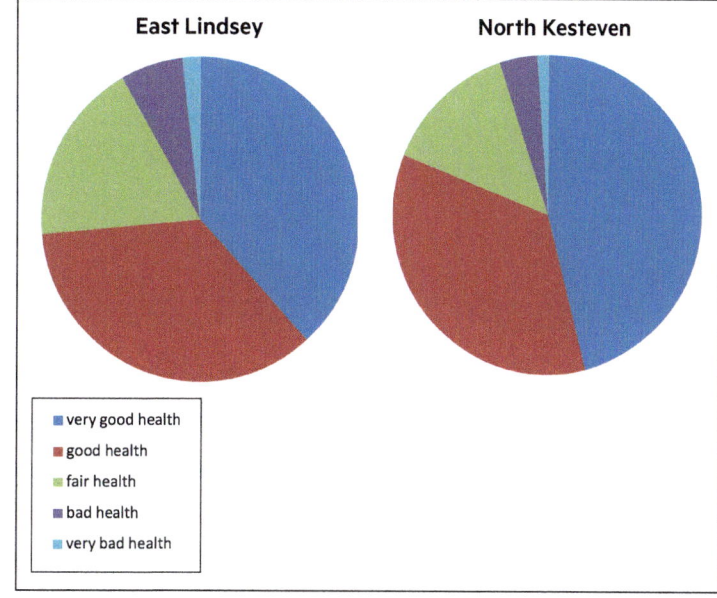

**Figure 15** Census data on health in two areas of Lincolnshire. East Lindsey is the area around Skegness. North Kesteven is a rural area south of Lincoln.

**Figure 16** A student has used the GIS map to plan this transect through rural Lincolnshire.

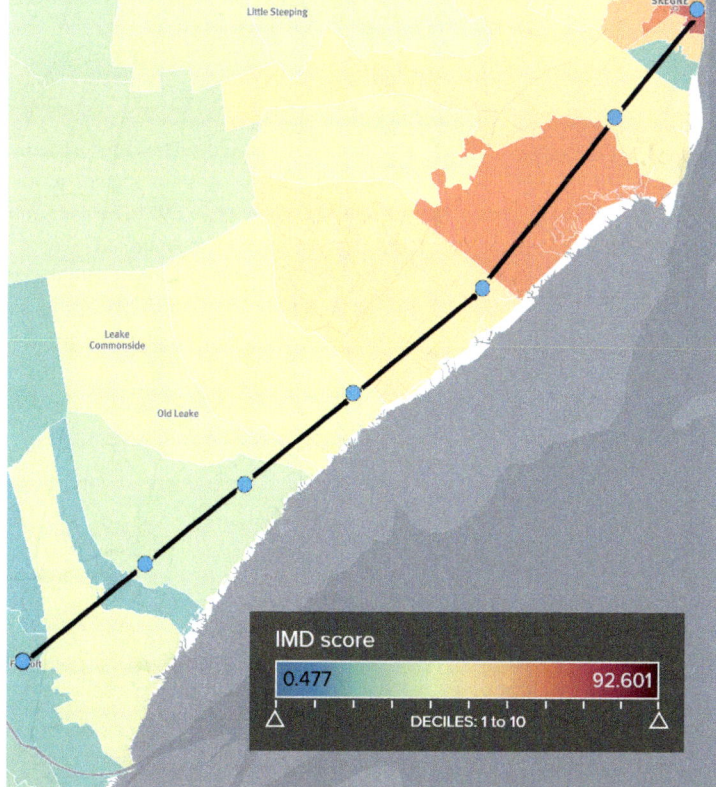

## Activities

1 Study Figure 15. Compare the health of the two areas.

2 A student used the GIS shown in Figure 14 to plan a transect. The route of the transect is shown in Figure 16.
   a) Suggest one advantage of using the GIS to plan a sampling strategy.
   b) Assess the design of the sampling strategy used by the student. Use evidence from Figure 16.

# Time to reflect

**Figure 1** The entrance to Birmingham's New Street train station.

1 **A student investigated sustainable urban living in the area shown in Figure 1.**
   a) Suggest one hypothesis or enquiry question that could be investigated here.
2 **The student designed an EQI to assess sustainable features of the city centre. The student's EQI is shown in Figure 2.**
   a) Identify **two** possible problems with the student's EQI in Figure 2.
   b) Suggest **one** way that this EQI could be improved.
   c) Suggest **one** way that data on the location of CCTV could be presented. Justify your choice.

**Figure 2** The student's EQI.

| Feature | Criteria | | | | |
|---|---|---|---|---|---|
| | **5** | **4** | **3** | **2** | **1** |
| Strong communities | Frequent CCTV cameras and a wide range of other crime prevention strategies are used. | Some CCTV cameras and at least one other type of deterrent, e.g. neighbourhood watch signs. | A lot of graffiti and some vandalism. | Good street lighting. | No evidence of any crime prevention strategies. |
| | **5** | **4** | **3** | **2** | **1** |
| Transport | Frequent bus services with bus lanes. Cycle lanes are well-used. | Lots of bus lanes and bus stops. | Some bus services and some cycle lanes. | Infrequent bus services. No cycle lanes. | No public transport. No cycle lanes. |

Pedestrian flows at 8:30 am

20 people

10 people

0   100m

approx scale

BUTE PARK

TAFF

RIVER

busy road

To the city centre

PRINCIPALITY STADIUM

To the train station

Pedestrian flows at 4:30 pm

20 people

10 people

0   100m

approx scale

BUTE PARK

TAFF

busy road

To the city centre

RIVER

To the train station

PRINCIPALITY STADIUM

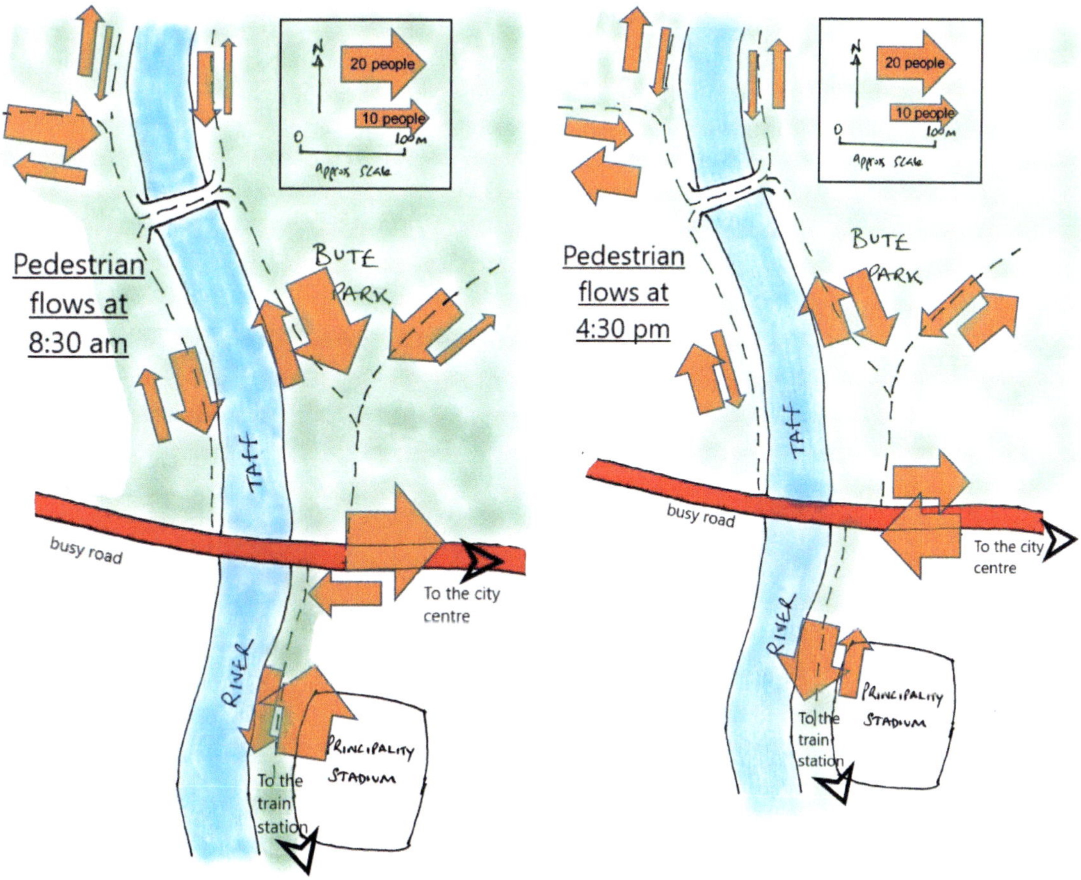

Figure 4 Students' flow line maps of pedestrian flows.

3  A group of students counted pedestrians walking through green areas of Cardiff. They split into 7 different groups to collect data at each of the locations shown in Figure 4.

   a) Identify **one** possible risk in the location shown in Figure 3.

   b) Identify **two** different reasons why their pedestrian counts may have been unreliable.

   c) Suggest **one** way of improving the reliability of their data.

   d) Explain **one** conclusion that can be drawn from Figure 4.

4  **Reflect on your own fieldwork in a human geography environment.**

Consider whether your data presentation was suitable and effective.

Part 02 Fieldwork in human environments

# Chapter 9

# Investigating coastal environments

## Learning objectives
- How to draw a beach profile.

## Investigating beach profiles

Some of the sediment on a beach is transported up the beach in the swash each time a wave breaks. It is then transported back down the beach by the backwash. The movement of the waves creates a series of slopes on the beach – the **beach profile** – sometimes with distinct ridges near the top of the beach. The angle of the slope depends on the size of the beach sediment. Sandy beaches tend to rest at gentle angles – usually less than 10 degrees – whereas pebble beaches usually rest at steeper angles. You can investigate a beach profile to see whether the steeper ridges coincide with larger pebbles.

**Figure 1** Swash and backwash on a pebble beach. Llantwit Major, Wales.

## Drawing a beach profile

To record a beach profile you will need a **clinometer** to measure the slope angles (see pages 20-21). You will need a data collection sheet to record the cumulative distance along the beach profile. Use it to record the angles and distances between each **break in slope** – an example is shown in Figure 2. It's a good idea to make a **field sketch** of the slope too – that will make it easier to interpret your data when you get back to the classroom. An example is shown in Figure 3.

|  | Angle (degrees) | Distance (metres) | Cumulative distance (metres) |
|---|---|---|---|
| Profile between start and break in slope 1 | 5 | 24 | 24 |
| Profile between break in slope 1 and 2 | 16 | 8 | 32 |
| Profile between break in slope 2 and 3 | 6 | 14 | 46 |
| Profile between break in slope 3 and 4 | 30 | 4 | 50 |
| Profile between break in slope 4 and end | 10 | 10 | 60 |

**Figure 2** Data collection sheet needed to draw a beach profile.

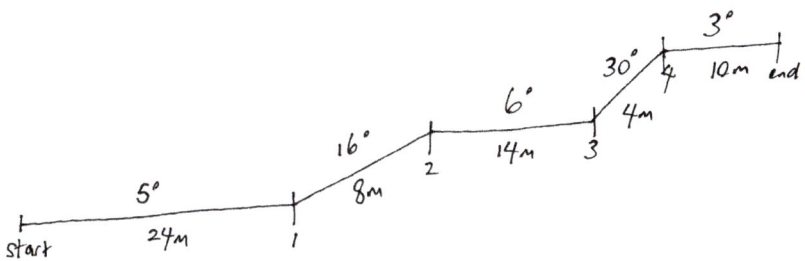

**Figure 3** A field sketch with the measurements and angles will help you visualise the slope later.

With the slope **data** you can represent the beach profile. Your finished drawing will be a **cross section** rather than a graph. You will need a ruler and protractor.

**Step One** Draw a horizontal line and work out a suitable scale to represent the cumulative length of the profile. This is the base line of your cross section.

**Step Two** Working from the left, use the protractor to draw a pencil line at the correct angle for your first reading (5 degrees in Figure 2) from the start of your base line.

**Step Three** Identify the first break in slope on the base line (24m in Figure 2). Place the centre of your protractor above this point, parallel to the base line, and mark off the second angle (16 degrees). Figure 4 shows the protractor being used at the third break in slope.

**Step Four** Continue to mark in all of your angles in pencil. When you have finished you can go over the finished profile in ink.

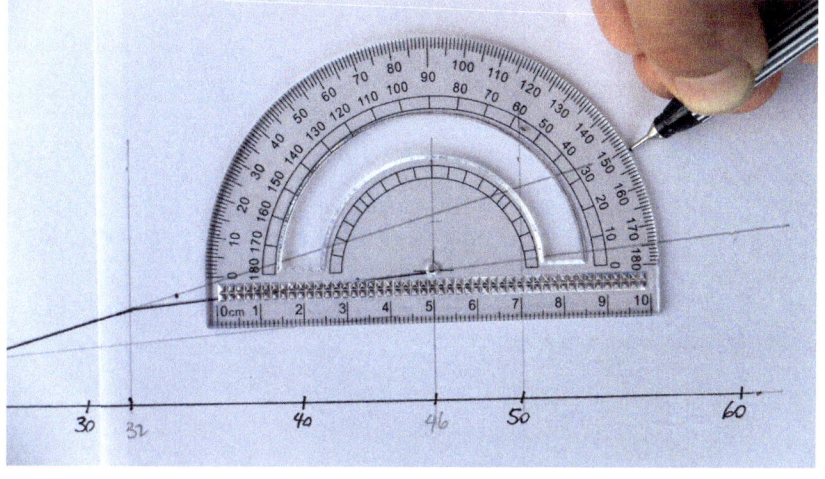

**Figure 4** Using a protractor to draw the beach profile.

# Histograms and bar charts

**Learning objectives**

■ Understand the differences between bar charts and histograms.

**Figure 5** The difference between bar charts and histograms.

A **histogram** is a type of graph that uses bars. It may look similar to a **bar chart** but there are some important differences between histograms and bar charts that are shown in Figure 5. Of these differences, the key one is that histograms are always used to represent **continuous data**. In geography **fieldwork**, the most common reason to use a histogram would be to represent pebble sizes from a beach, or the sediment sizes sampled from a river.

### Histograms

■ Represent **continuous data** (something that you have measured). This could be the length of pebbles, the height of plants, or the age of people.
■ The horizontal **axis** is a scale – just like in a **line graph**. The units of **measurement** (for example, centimetres) must be shown.
■ The bars should touch each other.

### Bar charts

■ Represent **discrete data** (something you have counted) such as types of vehicle in a traffic survey.
■ The horizontal axis is used to represent the different categories (for example, cars, lorries, and buses).
■ There should be a gap separating the bar for each category of data.

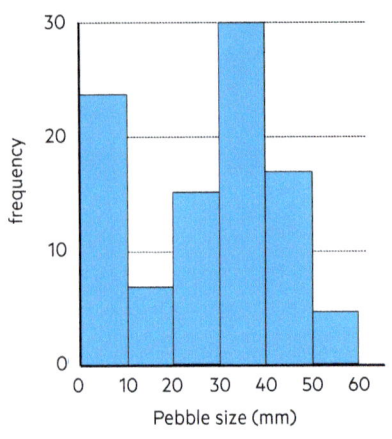

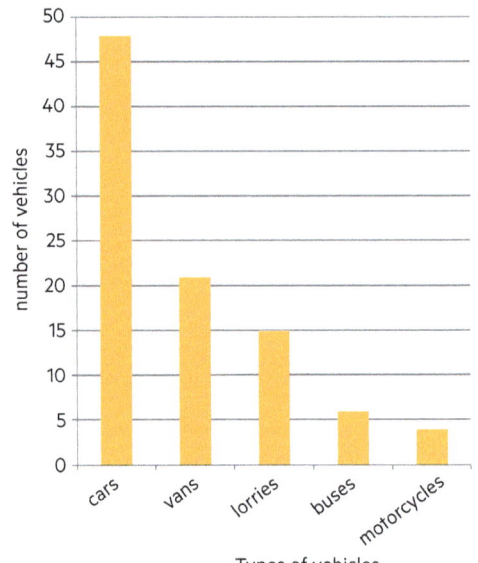

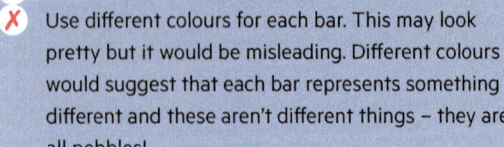

---

### Dos and don'ts of drawing histograms

**Do:**
✓ Ensure that the intervals on the horizontal axis are equal in size. This means each bar must be the same width.
✓ Use the same scale of horizontal and vertical axis for each of the histograms so you can compare them.

**Don't:**
✗ Use unequal sized intervals.
✗ Use different colours for each bar. This may look pretty but it would be misleading. Different colours would suggest that each bar represents something different and these aren't different things – they are all pebbles!
✗ Leave a gap between each bar.

# Investigating pebble size

Sediment tends to be sorted (by size) by the movement of the waves. Swash is what we call the movement of water up the beach as the wave breaks. Swash can provide a powerful force, moving sand, gravel, and pebbles of all sizes up the slope of the beach profile. The backwash usually has less energy. It drags smaller sediment back down the **beach profile** but larger pebbles are left behind higher up the slope. These larger pebbles sometimes form distinctive ridges near the top of the beach profile.

To investigate whether pebbles have been sorted by size you could start with a **hypothesis** such as:

## *'Pebble sizes are larger at the top of the beach profile than at the bottom of the beach profile.'*

To investigate this hypothesis you would follow these steps.

**Step One** Use a **clinometer** and tape to measure the gradient of the beach profile.

**Step Two** Collect data from 5 equally spaced sites along the beach profile. This would be a form of **systematic sampling**.

**Step Three** Measure the length of the same number of pebbles at each site. Do this using a **quadrat** and a set of **random number tables**.

**Step Four** Use your data to draw a histogram of pebble sizes at each sample point.

**Figure 6** Each pebble needs to be measured in the same way.

## Measuring pebbles

Pebbles are 3-dimensional objects with three axes – each **axis** is at right angles to the others. To record pebble size and shape you need to **measure** all three axes as shown in Figure 6. If you just want to record the size of the pebbles you can:

- measure the length of the b-axis only;
- measure all three axes for each pebble and then divide by 3 to get an **average**.

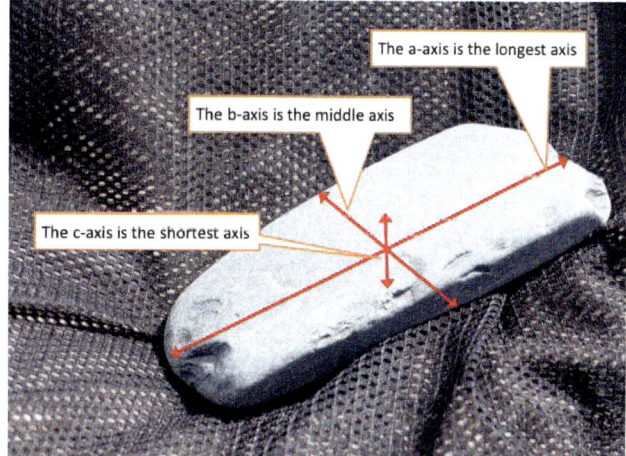

The a-axis is the longest axis

The b-axis is the middle axis

The c-axis is the shortest axis

## Activities

1 **Imagine you are going to investigate pebble sizes on a beach profile.**
   a) Explain why you should collect data:
      i) when the tide is going out;
      ii) when a large part of the beach is exposed.
   b) Suggest why it's a good idea to:
      i) choose sites systematically rather than randomly;
      ii) choose 5 sites rather than 2 (one at the top and one at the bottom of the beach).

# Investigating longshore drift

Sediment is transported along the coast by the process of longshore drift. As the waves and currents slow down they lose energy. As they do so they begin to deposit the sediment. The largest, heaviest sediment is dropped first. Smaller and lighter sediment is transported further and deposited later. This process is known as sorting. This means that beaches may have cobbles or pebbles at one end and smaller shingle or sand at the other.

**Figure 7** The process of longshore drift moves sediment from south to north in this diagram. Each groyne traps some sediment on its south facing side.

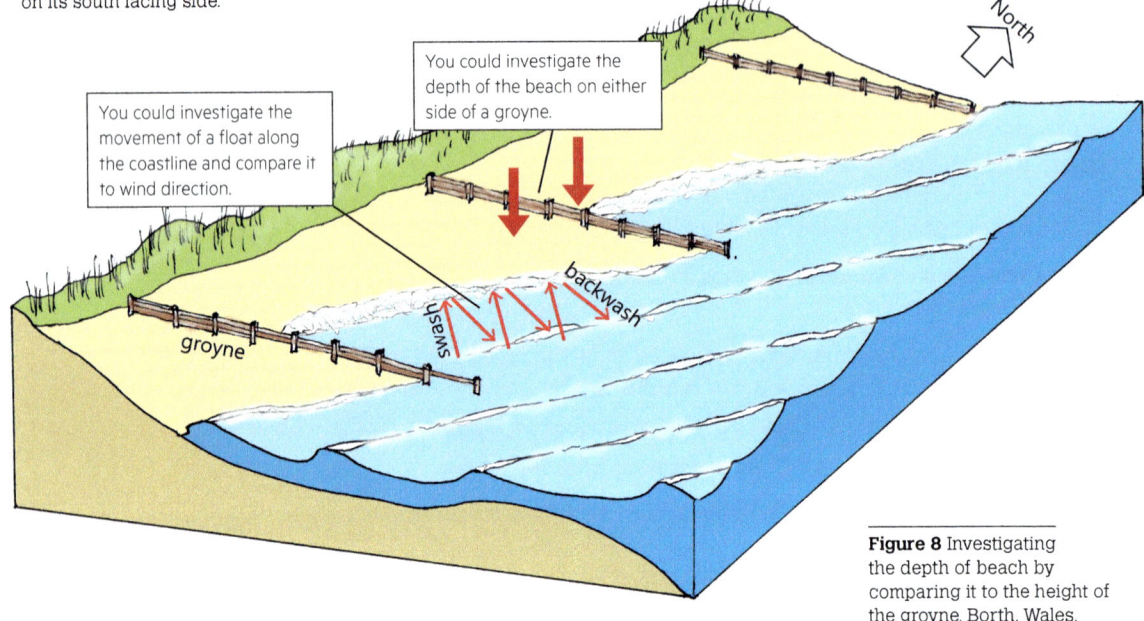

You could investigate the movement of a float along the coastline and compare it to wind direction.

You could investigate the depth of the beach on either side of a groyne.

North

backwash

swash

groyne

**Figure 8** Investigating the depth of beach by comparing it to the height of the groyne. Borth, Wales.

We could investigate the processes of longshore drift on a beach by:

- using a float to identify the direction and estimate the speed of longshore drift (see Figure 10);
- using **quadrats** to sample beach sediment at regular intervals along the beach;
- **measuring** the height of the beach on either side of each groyne (see Figure 8);
- measuring wind speed (using an **anemometer**) and wind direction and comparing this to the direction of longshore drift).

## Investigating groynes

Groynes are artificial structures made of heavy wooden boards or boulders. They are constructed on beaches to interrupt the movement of longshore drift. Groynes trap sand, shingle, and pebbles, making the beach thicker and wider. The beach is then able to absorb more wave energy, protecting the coast behind the beach.

You can investigate whether the groynes are trapping sediment by measuring down from the groyne to the surface of the beach. Do this on each side of the groyne – if it is trapping sediment then the beach will be thicker (and the distance smaller) on the side of the groyne that faces into the direction of longshore drift.

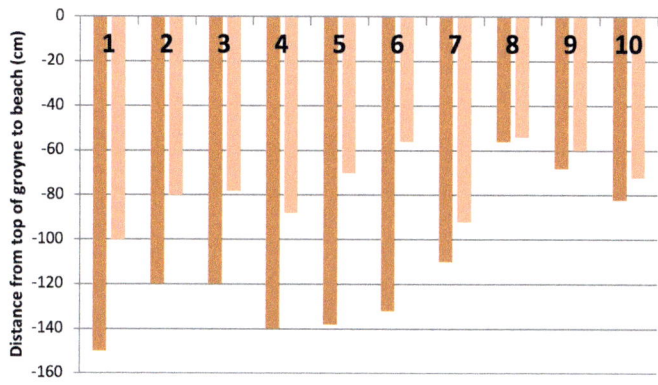

■ South side of groyne

■ North side of groyne

## Measuring speed of transport of material in the sea

You can use a float such as an orange or a tennis ball to measure the speed and direction of longshore drift. Figure 10 shows how to do this. For safety, you should only attempt this technique when the tide is going out.

**Step One** Measure a distance of 10 metres along the beach, as close to the water's edge as is safe. Mark each end.

**Step Two** Place a float in the sea. Don't throw it as the momentum of your throw might make the result **unreliable**.

**Step Three** Record the length of time it takes for the float to be washed by the swash and backwash along the beach for the 10 metres that you measured out.

**Step Four** Repeat the process at least three times. Then calculate the **average** time it takes for the float to travel 10 metres along the coast.

**Step Five** Calculate the **velocity** of longshore drift. Divide the distance (metres) by the average time (seconds) to get a velocity (metres per second).

Figure 10 How to measure the speed and direction of longshore drift.

## Activities

1 **Study Figure 9.**
 a) Evaluate this presentation technique.
 b) Suggest an alternative way of presenting the data. Explain why this might work better.
 c) Analyse this data. What conclusions can you draw about the direction of longshore drift?

# Investigating coastal management

The risk of coastal flooding is high in some parts of the UK such as the Norfolk coastline shown in Figures 11 and 13. However, coastal management is expensive so we need to **assess** whether coastlines are worth protecting. We could investigate the potential impacts of coastal flooding using a weighted **Environmental Quality Index (EQI)** like the one shown in Figure 12. In this EQI the **weightings** give larger scores for economic flood risks but you could alter these weightings to give the social and environmental risks greater importance.

**Figure 11** The metal flood gates (on the left of the photo) and a large earth embankment, at Sea Palling, protect homes and businesses from coastal flooding in the flat, low-lying coastal areas of Norfolk.

**Figure 12** A weighted EQI to help assess flood risk.

| | | | | | | | |
|---|---|---|---|---|---|---|---|
| **Economic** | Businesses at risk | Large shops or factories at risk | 40 | Small shops at risk | 20 | No businesses at risk | 0 |
| | Transport links at risk | Major roads or rail lines at risk | 30 | Minor roads at risk | 15 | No roads at risk | 0 |
| **Environmental** | Damage to habitats | Rare habitats at risk | 10 | Mixture of habitats at risk | 5 | Some low value habitats at risk | 0 |
| | Damage to buildings | Historic buildings at risk | 10 | Ordinary buildings at risk | 5 | No buildings at risk | 0 |
| **Social** | Risk to community | Vital services at risk | 20 | Non-vital services at risk | 10 | No services at risk | 0 |
| | Risk to residents | Large population including elderly | 30 | Small population | 15 | No local residents | 0 |

## Hard engineering strategies

Coastal management strategies include hard engineering strategies such as building sea walls, groynes, or the use of rock armour (large boulders). The impacts of hard engineering are usually positive (for example, protection from flooding) but they might also be negative. For example, sea walls can make access to the beach more difficult and rock armour could be perceived as ugly, or a potential danger for young children.

Groynes prevent longshore drift from transporting sediment along the coast and, in some cases, this has increased the rate of erosion elsewhere.

We could investigate hard engineering strategies by:

- taking and then annotating photographs like Figure 13;
- conducting **bipolar surveys** like Figure 14;
- asking local people about their attitude to coastal management in a **questionnaire**.

Study Figure 13. The **annotations** describe how the coastal management strategy protects the low-lying land behind. Notice how the annotations are numbered so that they tell a story.

**Figure 13** Use annotations to analyse how the management strategy works. Hard engineering at Sea Palling. See also page 33 for another photo of Sea Palling's hard engineering.

**1 Artificial reef.** These large boulders have been dumped below low tide to absorb the energy of waves driven towards the coast by an easterly wind.

**2 Beach.** Sand has been deposited behind the reef because the waves have less energy. This means the beach is now wider so it will absorb a lot more wave energy than a narrow beach.

| Positive evaluation | +2 | +1 | 0 | -1 | -2 | Negative evaluation |
|---|---|---|---|---|---|---|
| Good access to beach | | | | | | Poor access to beach |
| Likely to prevent floods | | | | | | Unlikely to prevent floods |
| Protects habitats | | | | | | Disturbs habitats |

**Figure 14** A bipolar survey designed to assess hard engineering strategies.

## Activities

1 **Study Figure 14.**
   a) Suggest another pair of evaluation statements that could be added to Figure 14.
   b) Explain why the use of Figure 14 by several different groups of students could give unreliable data.
   c) Suggest how you could make this technique more reliable.

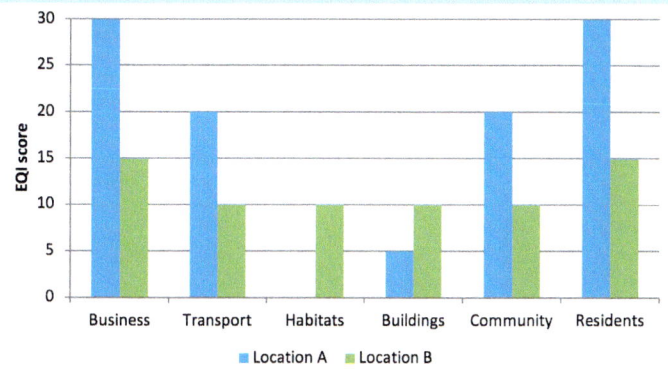

**Figure 15** EQI scores for two coastal locations.

2 **A student used Figure 12 to collect EQI scores in two coastal locations. One location was a ferry port. The other was a small coastal village. The student drew Figure 15 with the results.**
   a) Explain two conclusions that can be reached from this evidence.
   b) Assess the decision to weight the EQI in this way. Would you have done it differently? If so, why?

# Investigating sand dunes

Your case study of a coastal landscape may include an area of sand dunes. Sand dunes are a common feature of coastal landscapes where deposition is occurring. They often form on spits and at the top of sandy beaches beyond the high tide mark. Sand dunes are formed when sand is blown from the beach. The sand is transported up the beach and then deposited around the salt-tolerant plants that grow at the top of the beach.

**Embryo dunes** A few specialised plants that are tolerant of salt spray live here, like this sea rocket. They take nutrients from decomposing seaweed washed up in the strand line near the top of the beach.

**Mobile dunes** Marram grass grows here – it has large fibrous roots that hold the sand together. The tall leaves stick up into the air and slow the wind speed so that sand carried in the wind is deposited. Wind easily erodes sand where there are no plants.

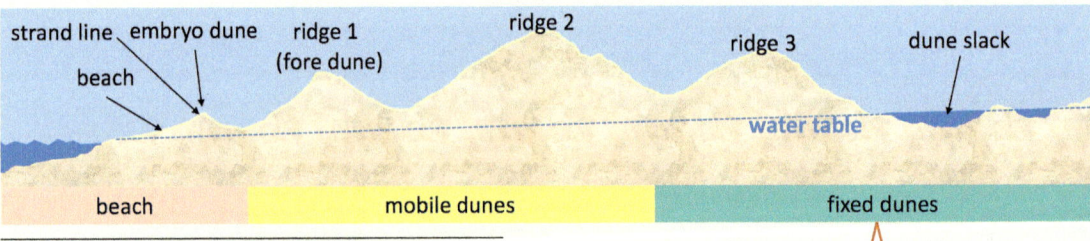

**Figure 16** A cross section through a sand dune ecosystem.

You can create your own sand dune cross section by extending your beach profile beyond the strand line and into the dunes. You will need a tape and clinometer as described on page 97.

**Fixed dunes** This is the oldest part of the dunes. Wind speeds are lower here so erosion is less likely. There are more nutrients in the soil because plants have been dropping leaf litter here for longer than elsewhere in the dunes. Consequently, the fixed dunes have a wide variety of flowering plants, shorter grasses, and shrubs like this bramble.

# Management

Sand dunes are vulnerable to erosion so they need management to keep their ecosystem healthy. Where people trample through the dunes the marram dies and the sand is eroded by the wind. If the sand dunes are badly damaged they are at risk of erosion during storms. Management techniques include:

- fencing off areas that have been eroded to prevent further damage and replanting with marram;
- the creation of boardwalks to control the movement of people through the dunes;
- information boards and signage to educate visitors.

Sand dunes absorb wave energy during a storm and protect the land behind from flooding. The management of sand dunes so that they are healthy is an important part of managing the coastline. Using a natural method to manage the coast is called soft-engineering.

https://www.rgs.org/schools/teaching-resources/ecosystems/
The RGS website gives more details about fieldwork in sand dunes.

## Activities

1 Study Figure 16. Students recorded plant types along the cross section. They used stratified sampling to record wind speed on each ridge and in the trough between each ridge.
   a) Identify **one** enquiry question or hypothesis that could be investigated using fieldwork in this environment.
   b) Suggest why stratified sampling could allow you to collect data more quickly than systematic sampling if you need data from each feature in the sand dunes. Use Figure 16 to help you.
2 Study Figure 17. It shows the results of the students' transect through the dunes.

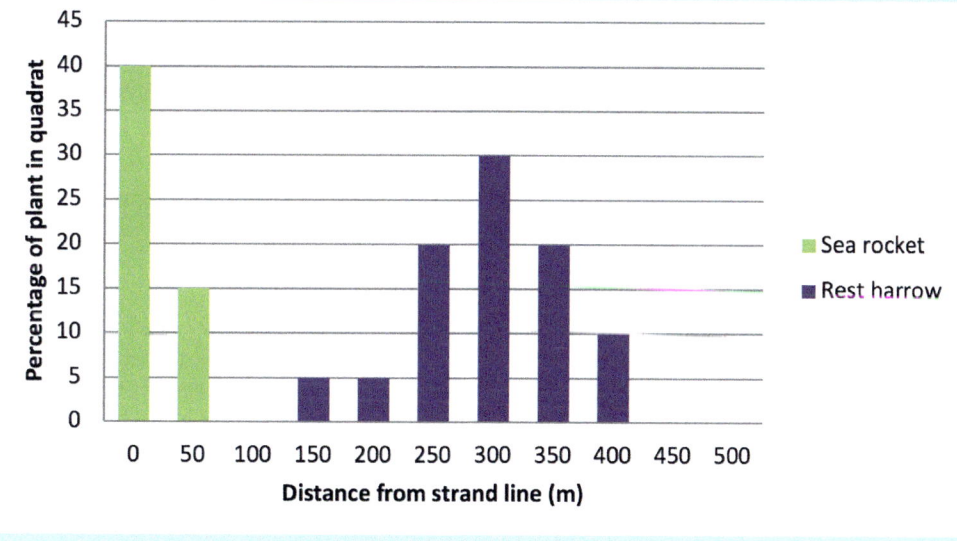

   a) Describe the pattern that can be seen in Figure 17.
   b) A student had the following hypothesis:
   *'Sea rocket only grows in embryo dunes'.*
   To what extent does the evidence in Figure 17 prove this hypothesis to be true?
3 Imagine your aim is to investigate the effectiveness of sand dune management.
   a) Decide what data you would need to collect.
   b) Describe how you would collect it.
   c) Explain why the data collection technique you have chosen would be suitable and provide reliable data.

# Time to reflect

Figure 1

**1 Study Figure 1. It shows a possible fieldwork location.**

**Figure 2** Data from Site 1. Measurements along the middle axis of each pebble.

| Pebble | b-axis (cm) |
|--------|-------------|
| 1 | 18 |
| 2 | 2 |
| 3 | 1 |
| 4 | 6 |
| 5 | 4 |
| 6 | 8 |
| 7 | 9 |
| 8 | 2 |
| 9 | 11 |
| 10 | 2 |
| 11 | 3 |

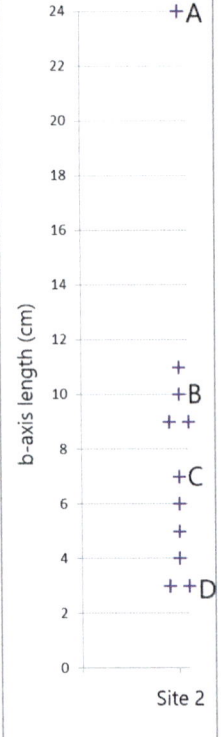

a) Identify **one** potential risk that could occur while doing fieldwork in this environment.

b) Suggest **one** way that this risk could be reduced.

c) Identify one hypothesis that could be investigated here.

d) Identify **two** data collection techniques that could be used to investigate this environment.

e) Use the evidence in Figure 1 to complete this annotation by choosing answer i), ii), or iii).

There is evidence that swash has ...

   i) eroded larger pebbles from the upper beach.

   ii) rolled larger pebbles up the beach.

   iii) deposited sand on the upper beach.

**2 A student measured pebble sizes at two sites: one at the bottom of the beach and the other at the top of the beach. Data for Site 1 is shown in Figure 2. Data from Site 2 is shown in Figure 3.**

**Figure 3** Data from Site 2 represented in a dispersion graph.

a) Give the median pebble size at site 1.

b) Explain why the **median** is a more useful measure than mean for this data.

c) Calculate the **interquartile range** at site 1. Show your working.

d) Match the following features to labels A-D on Figure 3.

**Upper quartile**    **Maximum**    **Minimum**    **Median**

e) Calculate the range at site 2. Show your working.

f) The students came to the conclusion that pebbles at the top of the beach are larger than pebbles at the bottom of the beach. Explain why this conclusion may be unreliable.

3  A group of students measured pebbles at five sites on a beach profile. Their results are shown in Figure 4.

**Figure 4** A student's photo and histograms.

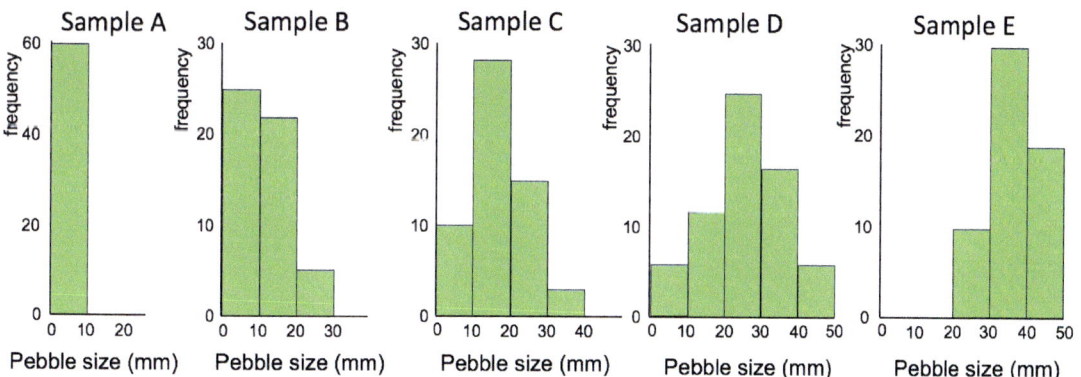

a) Suggest a suitable annotation that explains the evidence at A in Figure 4.

b) What conclusions can you reach from the evidence in Figure 4?

**4  Reflect on your own fieldwork in a physical geography environment.**

a) Did you use a sampling technique? What were its strengths and weaknesses? Can you justify the method that you used?

b) Consider how you collected primary data. Do you think this method gave you data that was reliable?

c) Did you use any secondary data? If so, how useful was this data?

# Chapter 10

# Investigating river environments

## Learning objectives

- How to draw a cross section of a river channel.

# Investigating river processes

The processes of erosion, transportation, and deposition all happen in river channels like the one shown in Figure 1. Sediment is transported down the river. Where water has less energy, as in the shallow water at X, sediment is deposited. You can see **evidence** of this in the photograph – pebbles have been deposited in a river beach, also known as a slip-off slope or point bar. Erosion also occurs in meandering river channels. Erosion tends to occur on the outside bend of the meander where water flows with greater energy - forming a river cliff like the one at Y in Figure 1.

**Figure 1** A meandering river channel.

We could investigate the processes in this river channel by:

- **measuring** the depth and width of the river channel and drawing a **cross section**;
- measuring the velocity of the water and calculating **discharge** (see pages 112-3);
- **analysing** the size of pebbles on the river's bed and slip-off slope to find evidence of sorting;
- taking photos or making **field sketches** and **annotating** the main features (see pages 32-34).

## Creating a cross section of a river channel

To create a **cross section** of the river channel you will need to follow these steps.

**Step One** Measure the width of the river channel. Divide this by 10 to create 10 evenly spaced sample points. In Figure 2 the channel is 3m across, so the samples are 30cm apart.

**Step Two** Stretch a line tightly across the river. Make sure it is level.

**Step Three** Measure downwards, from the line to the ground, at each sample point. In Figure 2, measurements were made every 30cm on each side of the river as well as into the river channel. Record the distance from the line to the surface of the water.

**Figure 2** How to take data readings to create a cross section across a river channel. Shropshire Hills, Area of Outstanding Natural Beauty (AONB).

| **Dos and don'ts of river cross sections** | |
|---|---|
| **Do:** | **Don't:** |
| ✓ Include some of the river bank on either side so that you can draw the river's banks on your cross section. Then you can estimate how much water would be in the river channel when river levels rise. | ✗ Let the line droop in the middle. If it does, the measurements for the middle of the channel will be too small. |

## Drawing the river's cross section

To draw the river channel's cross section you need to follow these steps.

**Step One** Draw a base line (or **x-axis**) that represents the width of the river .

**Step Two** Add a **y-axis** to represent the measurements down from the tape measure to the stream's bed.

The finished graph will look something like Figure 3. Notice that the first and last measurements, in Figure 3, were at the very edge of the stream. The student has also added a horizontal line and some colour to represent the depth of the water.

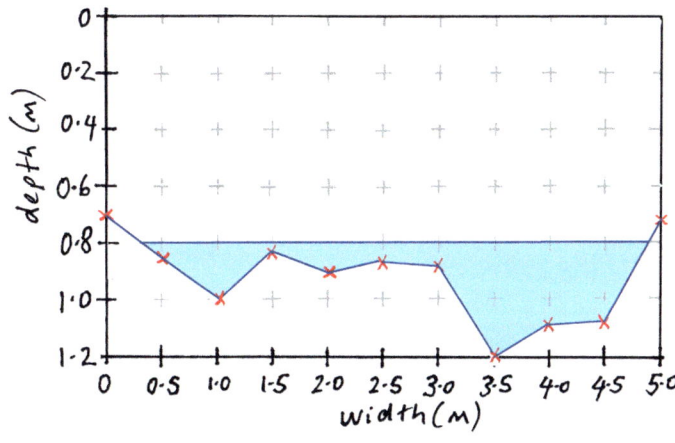

**Figure 3** A student has used the technique shown in Figure 2 to collect data for a cross section.

## Activities

1 **Study Figure 1.**
   a) Identify **two** potential risks that could occur while doing fieldwork in this environment.
   b) Discuss how each of these risks could be reduced.
2 **Study Figure 4. It shows two common mistakes made by students when drawing a cross section of a river. Identify the errors.**

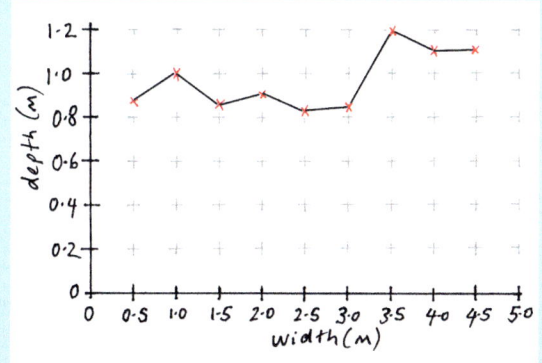

**Figure 4** An incorrectly drawn cross section. This graph has been drawn using the same data as Figure 3.

Part 03 Fieldwork in physical environments

# Investigating slip-off slopes

Slip-off slopes (point-bars) are common features in meandering river channels. They form where water flows more slowly on the inside bend of the meander. Large pebbles can be transported along the river's bed in the deepest part of the channel where the river has plenty of energy. Higher up the slip-off slope the water flows slowly so pebbles are deposited. To investigate whether pebble sizes vary up and down a slip-off slope, use a **hypothesis** such as:

### 'Pebble sizes are smaller at the top of the slip-off slope than in the deepest part of the river channel.'

To investigate this hypothesis you would need to follow these steps.

**Step One** Set up a **transect** at right angles to the river and up the slip-off slope.

**Step Two** Decide on a **sampling strategy**. For example, you could collect data from 10 equally spaced sites along the transect as in Figure 5.

**Step Three** Select and measure the same number of pebbles from each sample site. Measuring the b-axis is the quickest way to do this (see page 98).

**Step Four** Draw a **histogram** (see page 98) to **analyse** your results.

**Figure 5** Regular sampling of pebbles from the deepest part of the river channel to the top of the slip-off slope.

## Investigating downstream changes in sediment

Rivers go through major changes as they flow from source to mouth. Not all rivers behave in the same way, but for many rivers in the UK:
- the river channel gets wider and deeper. This means that there is proportionally less water in contact with the river bed so there is less friction. This means the **velocity** of the water increases;
- sediment in the river channel gets smaller as you go further downstream. It also gets more rounded. This is because the pebbles have been worn down for longer by the process of attrition.

We could investigate how sediment changes by visiting several sites along a river and recording:

- the length of the b-axis in a **sample** of 25 pebbles from the stream's bed;
- whether the pebbles are angular (**evidence** of little attrition) or rounded (evidence of lots of attrition). We can do this by comparing each pebble in the sample to a roundness scale like the one shown in Figure 6.

1 Very angular  2 Angular  3 Sub-angular

4 Sub-rounded  5 Rounded  6 Well-rounded

**Figure 6** A six point scale of roundness can be used to categorise each pebble in your sample.

## Activities

1 **Study Figure 7. Site 1 was close to the source of the stream. The sites were 10km apart.**
   a) i) How many pebbles were angular at Site 1?
      ii) What percentage of pebbles were angular at site 1?
   b) What conclusions can you reach from the evidence in Figure 7?
   c) Explain why it might be very difficult to collect data from a river at exactly 10km intervals.

**Figure 7** A student's graph showing pebble roundness at three sites.

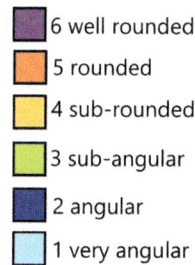

- 6 well rounded
- 5 rounded
- 4 sub-rounded
- 3 sub-angular
- 2 angular
- 1 very angular

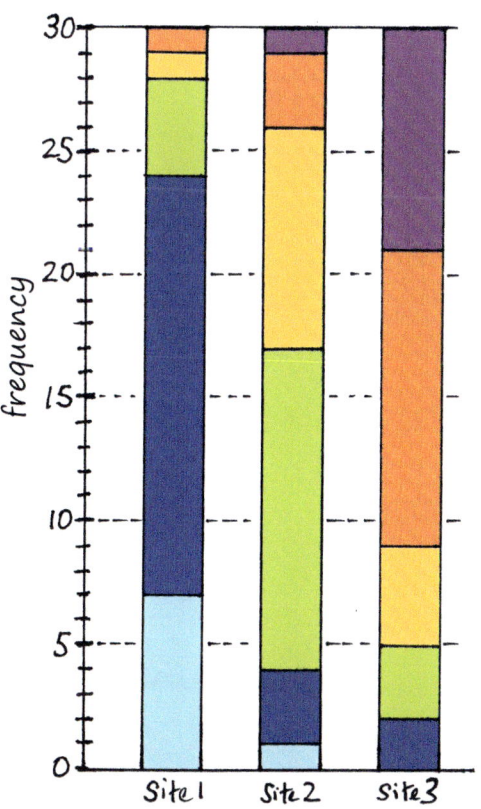

# Investigating river discharge

River **discharge** is the amount of water flowing down the river channel per second. Discharge is **measured** in cubic metres per second (cumecs). To investigate discharge you will need to:

- measure the width and **average** depth of the river channel. This will allow you to calculate the area of water in the river channel's **cross section**;
- measure the **velocity** of the river's flow.

One way to measure velocity is with a **flow meter**. You can see one being used in Figure 8. You should stand downstream of the flow meter. If you stand upstream your legs will disturb the flow of water and make the results inaccurate. The advantage of a flow meter is that you can use it to record velocities at precise depths in the water channel and also at different points across the width of the channel.

You can also calculate velocity by timing how long it takes for a float to travel downstream. Use something biodegradable like a dog biscuit as the float. You will need two people, a tape, and a stop watch.

**Step One** Measure a length of 10 metres along the river bank.

**Step Two** Place the dog biscuit in the water at the start of the 10 metres. Time how long it takes for the float to reach the end of the course.

**Step Three** Repeat the experiment at least three times. Calculate an average time by adding the results together and dividing by the number of experiments.

**Step Four** Calculate the velocity by dividing distance by the average time it took for the float to travel downstream. For example, if the average time was 78 seconds then the velocity would be 10 metres divided by 78 seconds, giving a stream velocity of 0.128 metres per second.

**Figure 8** Students measuring river flow with a flow meter.

## Calculating discharge

River discharge is calculated by multiplying the river **velocity** by the cross-sectional area of the channel.

In GCSE Geography, cross-sectional area can be calculated as:

width of the river channel (m) **x** average mean depth (m) = $m^2$

Then to calculate the river discharge:

cross-sectional area ($m^2$) **x** velocity (m/s) = discharge $m^3$/s (cumecs).

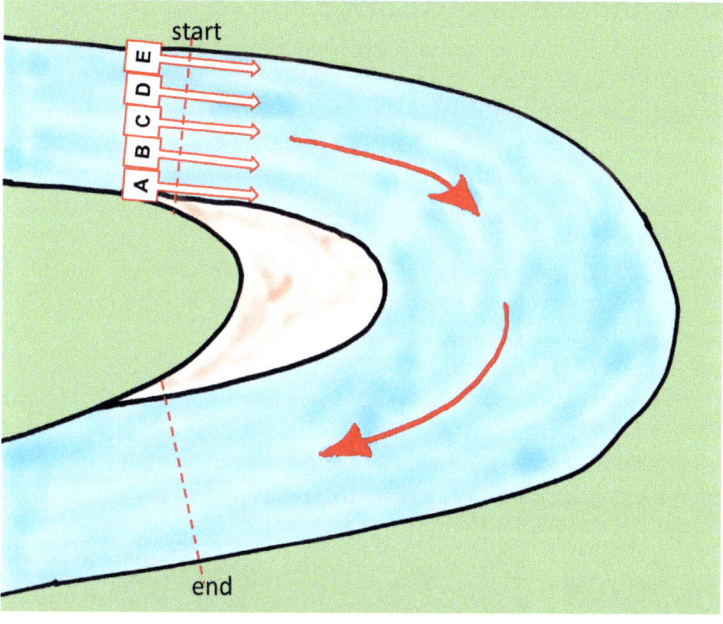

**Figure 10** Sketch map of the river in Figure 5 showing the five float positions and the start and end points used to measure the river's flow.

**Figure 11** Table of river velocity readings taken at the river shown in Figure 9.

| Float | Float time (seconds) | | | Total (seconds) | Average time (seconds) | Velocity |
|---|---|---|---|---|---|---|
| | First | Second | Third | | | |
| A | 54 | 50 | 58 | 162 | 54 | 0.19 |
| B | 57 | 61 | 59 | 177 | 59 | 0.17 |
| C | 64 | 98 | 66 | 228 | | 0.13 |
| D | 67 | 68 | 72 | 207 | 69 | |
| E | 78 | 82 | 89 | 249 | 83 | 0.12 |

## Activities

1  **Study Figures 9, 10, and 11. A student investigated the hypothesis:**
   *'Water flows faster on the outside bend of the meander'.*
   a) Suggest one strength of the sampling method used by the student.
   b) Identify one anomaly in this data.
   c) Calculate the average time for float C.
   d) Calculate the velocity for float D.
   e) What conclusion can you reach from studying the evidence in Figure 9?
   f) Use the evidence in Figure 11 to explain how the landform in Figure 9 is likely to change in the future.

# Investigating changes in discharge

A number of factors affect the amount of water (or **discharge**) flowing through a drainage basin. Rock type, land use, and the amount of precipitation are three important factors. We can collect primary **evidence** to investigate the location and extent of past flood events. We can use evidence collected from secondary sources to investigate how and why floods have occurred.

## Primary evidence of past floods

It is not safe to collect discharge **data** in a river that is in flood. However, we can observe evidence that tells us a lot about the extent of a past flood event by:

- looking for physical evidence that the river has flooded recently. This will take the form of rubbish – often plastics – caught in the branches of trees close to the river. Figure 12 shows an example. The height of this rubbish above the river bank can be measured with a tape to give an idea of the level of the flood water;
- the observation of features that indicate that the river has flooded in the past. These features can be photographed, annotated, and their position located on a **base map**. For example:
  - buildings with steps up to the front door so that the ground floor is above flood height – like in Figure 13;
  - place or street names that indicate a history of flooding;
  - flood defences such as embankments, floodgates, or fixings in the pavement for demountable defences to be slotted in.

**Figure 12** Rubbish in the branches of a tree next to the River Severn (Shrewsbury) in March 2016.

**Figure 13** This new house in Shrewsbury has been built above flood level. Notice the steps up to the front door - these tell us how high the river has flooded in the past.

# The importance of secondary data

Secondary sources can provide data that would be impossible to collect through **primary data** collection. In particular, **secondary data** allows you to investigate two important aspects of a drainage basin that would be very difficult, or time consuming, to do using only primary data:

- how discharge changes throughout the year. Websites that publish hydrographs, like Figure 14, can be used to identify flood events. If you can find rainfall data for the same period then you can investigate **correlations** between rainfall and discharge;

- **spatial** variations in geology and land use across a drainage basin. For example, you can use a geology app to see whether the catchment area of your river is mainly permeable or impermeable rocks. You can also use Google Earth to investigate the extent of different land uses in the drainage basin such as woodland, arable fields, and urban areas. Each of these land uses will influence how quickly a river's discharge changes after heavy rainfall - the so-called lag time. Woodland will intercept water so lag time is increased. Urban areas contain a lot of impermeable surfaces such as tarmac so run-off is increased and lag times are reduced.

https://www.riverlevels.uk/
Use this website to view hydrographs like Figure 14.

**Figure 14** A hydrograph for the River Severn in Shrewsbury (November 2015 – July 2016).

— Typical Low — Measurement — Typical High

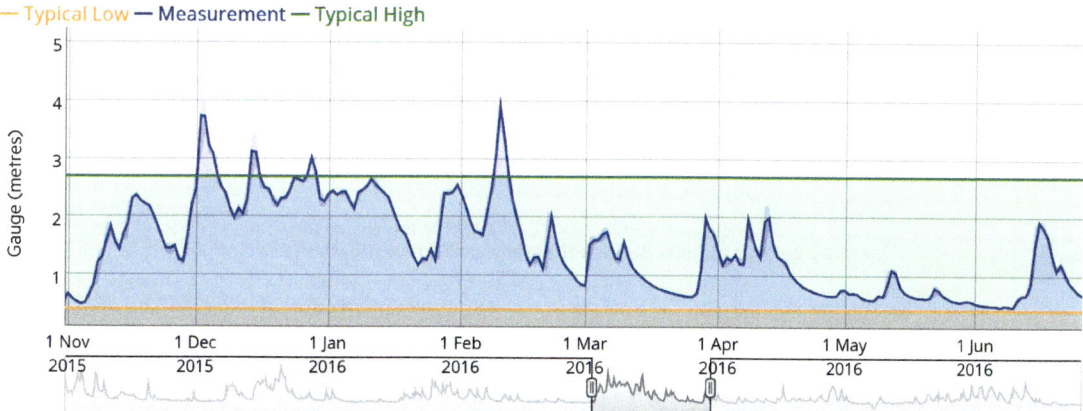

| Strengths and limitations of using secondary data in a river investigation | |
|---|---|
| **Some strengths** | **Some limitations** |
| • Secondary data such as satellite images allow you to investigate the whole drainage basin.<br>• It is possible to investigate variations in discharge over much longer periods of time than is possible with just primary data. | • Satellite images may be out of date and land uses may have changed.<br>• Discharge data for the river where you collected your primary data may not be available online. It may not be valid to use data from another nearby river. |

## Activities

1  **Study Figure 14. The orange line shows the typical depth (in metres) of the River Severn in Shrewsbury.**

   a) How deep was the river in mid-March, when Figure 12 was taken?

   b) When was the river at its deepest?

   c) Use Figure 14 to explain what can be seen in Figure 12.

# Time to reflect

**Figure 1** The Afon (river) Elan in mid-Wales.

1 **A group of students decided to investigate the location shown in Figure 1.**

   a) Identify one potential risk at this location.

   b) Explain how this risk could be reduced.

2 **The students decided that they would investigate the size of pebbles at this location.**

   a) Suggest a possible hypothesis or enquiry question that could be investigated at this location.

   b) Suggest how data could be collected at **this** location. Consider how you would ensure the data was accurate and reliable.

3 **Study Figure 2.**

   Suggest how this diagram should be redrawn so that it more accurately shows the shape of the river's cross section.

**Figure 2** A student's cross section of the river shown in Figure 1.

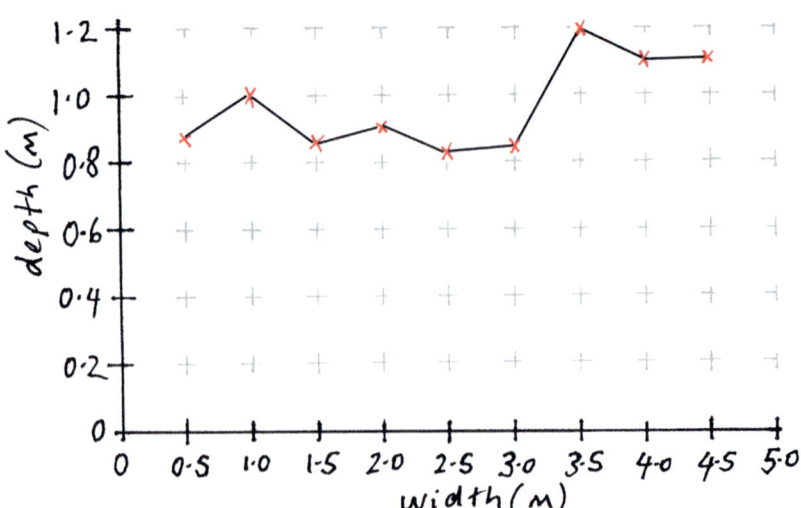

**4** **The students found a second site, 5km further downstream. It is shown in Figure 3.**

    a) Suggest a possible hypothesis or enquiry question that could be investigated using data from both sites.

Figure 3 The Afon Elan at Pont Elan.

**5** **The students took cross section measurements at three sites. Sites A and B were 5km apart. Site C was 3km further downstream. Figure 4 represents the three cross sections.**

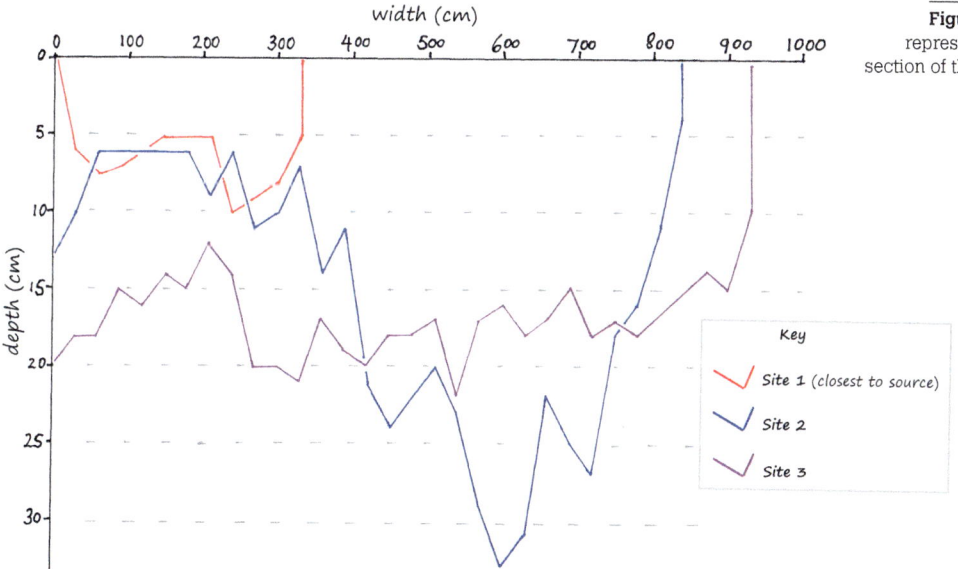

Figure 4 A diagram representing the cross section of the same river at three sites.

    a) Identify **one** weakness in the way the students chose the three sites.

    b) Study Figure 4. Compare the three cross sections using data from the graph to describe similarities and differences.

**6** **Reflect on your own fieldwork in a physical geography environment.**

    a) Consider what made the location suitable.

    b) Critically assess one presentation technique you used. Think about whether it was suitable, effective and accurate.

# PART 04

# Chapter 11

# Getting ready for the exam

## Learning objectives

- How fieldwork will be assessed.
- How to prepare for your fieldwork exam.

# How is fieldwork examined?

**Fieldwork** is assessed in an examination at the end of your course. The exam questions will assess you in two ways:

- first, whether you can interpret, analyse, and evaluate geographical information and make judgements about fieldwork;
- secondly, whether you understand geographical skills such as using maps and graphs and numeracy tasks such as the calculation of **mean** or **interquartile range**.

Questions could use any of the geographical skills listed in the specification. Geographical skills are described in this book on pages 40-55, 63, 77, 79, and 98.

Some questions in the exam will be about your own experience of fieldwork. Other questions will be based on some resources that you will see for the first time in the exam – questions about unfamiliar fieldwork.

## OCR A or OCR B?

You need to find out whether you are studying OCR A or OCR B. Look at Figure 1. It shows where fieldwork is examined in your specification.

**Figure 1** How fieldwork is assessed in OCR Geography.

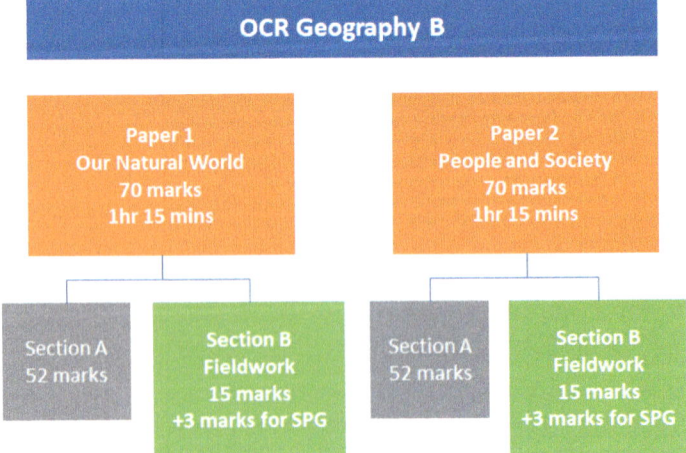

## OCR A

If you are studying OCR Geography A, your understanding of fieldwork is assessed in *Paper 3* (*Geographical skills*). Paper 3 is marked out of 80 marks. You will have one hour and thirty minutes to do the exam.

The exam paper is divided into **two** sections. The questions about fieldwork are in *Section B*. You must answer **all** of the questions in *Section B*. *Section B* is marked out of 33 marks which includes 3 marks for the accuracy of your spelling, punctuation, and grammar and your use of specialist terminology (SPaG).

- You will be asked questions about both human and physical geography **fieldwork**.
- The questions in *Section B* could be about any of the six stages of the **enquiry process** (see page 6).
- To answer some questions you will need to respond to stimulus materials in the exam paper. These stimulus materials could be photos of fieldwork sites, tables of data, or any of the maps and graphs described in this book (pages 40-55). Other questions will be about your own fieldwork.

## OCR B

If you are studying OCR Geography B, your understanding of fieldwork is assessed as part of *Paper 1* (*Our Natural World*) and also as part of *Paper 2* (*People and Society*). You will have one hour and fifteen minutes to do each of these exam papers. Each paper is marked out of 70 marks.

Each exam paper is divided into **two** sections. The questions about fieldwork are in *Section B* of each paper. Each *Section B* is marked out of 18 marks which includes 3 marks for the accuracy of your spelling, punctuation, and grammar and your use of specialist terminology (SPaG).

- You will be asked questions about physical geography fieldwork in *Section B*, *Paper 1* (*Our Natural World*).
- You will be asked questions about human geography fieldwork in *Section B*, *Paper 2* (*People and Society*).
- The questions in *Section B* of both papers could be about any of the six stages of the **enquiry process**.
- To answer some questions you will need to respond to stimulus materials in the exam paper. Other questions will be about your own fieldwork.

### What are the fieldwork questions like?

Fieldwork questions assess two things, your ability to:

- select, adapt, or use geographical skills, for example, selecting a suitable style of graph, suggesting how a graph could be adapted to improve it, or describing the pattern on a map;
- think critically about fieldwork by interpreting, analysing, or evaluating information, or making a judgement.

When you answer questions about your own fieldwork there are no marks for simply describing what you did. However, you may be asked to state the aim of your investigation and describe the location of your fieldwork. Some questions require you to think critically so it's useful to be able to use your own fieldwork experience to support any interpretation, analysis, evaluation or judgements that you make. This means that preparing for the fieldwork examination is a little different to how you might usually prepare for an exam. Pages 120-125 suggest a few strategies that you might like to try.

# How can you prepare?

## Learning objectives

- How to prepare for your fieldwork assessment.
- How to organise your fieldwork notes.
- How to evaluate how well you processed and presented the data for your fieldwork.

In order to get ready for the examination you should:

- **organise your fieldwork notes using six headings:** one for each stage of the **enquiry process**. Use Part 01 of this book to help you;
- **think about the main methods you used to collect data.** What were their strengths and weaknesses?
- **make sure you justify your decisions.** Can you explain why you used a particular sampling strategy, or presentation technique? What made it suitable? Keep notes about the decisions and choices you made;
- **use different techniques to process and present data.** Discuss which ones work best.

Fieldwork has its own list of specialist terms. You will find these terms in the glossary at the back of this book. You should make sure you are familiar with these terms so you can use them correctly when you are discussing your fieldwork investigations. Some particularly useful ones are listed in Figure 3.

**Figure 2** Fieldwork checklist.

| Task | I've done it |
|---|---|
| 1 I have **organised** my fieldwork portfolio and I have notes about each of the fieldwork enquiries. | |
| 2 I know the **title** for each of my fieldwork enquiries. | |
| 3 I remember the **aims** of each enquiry. | |
| 4 I understand the **sampling** methods we used and I can give at least one detailed reason why we used them. | |
| 5 I can **evaluate** the strengths **and** weaknesses of the data collection methods we used to give a **balanced** answer. | |
| 6 I can assess whether the maps and graphs that I drew were **suitable** and **effective** ways to represent my data. | |
| 7 I can analyse data and use this evidence to reach **conclusions** about my fieldwork. | |
| 8 I can discuss the **reliability** and **accuracy** of the data I used. | |
| 9 I can give **three** ways that my fieldwork could be **improved**. | |

**Figure 3** Key terms that can be used when evaluating fieldwork.

| | Term | Definition |
|---|---|---|
| Primary | | |
| Bias | | A tendency for evidence to be unreliable. |
| Transect | | Data that is counted as whole numbers or put into categories. |
| | | Collecting data from along a line. |
| Continuous | | |
| | | Data that can be measured to a decimal point. |
| Reliability | | Evidence that has already been published in another source. |
| Accuracy | | How close a measurement is to the true value of the object (or population) that is being measured. |
| Secondary | | Evidence that has been collected for the first time, by you or someone in your team. |
| Discrete | | Collecting data from regular intervals. |
| | | The degree to which the evidence collected during an investigation can be considered consistent and dependable. |
| Systematic | | |

## Activities

1 Make a copy of Figure 3. Add the correct term to each definition.
2 Make a copy of Figure 4. Use it to evaluate each stage of your enquiry process.

# Reflecting on your enquiries

The most useful way of preparing yourself for the fieldwork assessment is to think critically about what you did and why you did it during each of your enquiries. Remember, there are no marks for describing your fieldwork. Instead, think about each step of the **enquiry process** and **evaluate** what went well and what wasn't so good.

There are some techniques you can use to help with your evaluation. For example, you could use a fish bone diagram (page 124) or a Lotus Diagram (pages 122-123) to help you evaluate a particular part of your investigation. To get an overview of your whole fieldwork experience, you could make a large copy of Figure 4. Use the questions in Figure 4 to think about specific things that you did during your fieldwork that worked well or not so well. Fill as much of the table as you can with specific details of these strengths and weaknesses.

**Figure 4** You should reflect on each stage of your fieldwork enquiries.

| Stage of the enquiry process | Questions to ask yourself | Strengths of my fieldwork | Weakneses of my fieldwork |
|---|---|---|---|
| **1** Creating aims and selecting suitable questions | Did I have SMART aims? (See page 9.) Where did I decide to collect my data? Looking back, were these the best places? | | |
| **2** Selecting, measuring and recording data | Did I use systematic, random, stratified, or opportunistic sampling? Can I justify my choice? What might have happened if I had used a different sampling strategy? | | |
| **3** Processing and presenting fieldwork data | Did I collect discrete or continuous data? Did I choose the most suitable methods to represent this data? | | |
| **4** Analysing fieldwork data | Was I able to see patterns, trends, and correlations in my data? Would trends have been clearer if I had used more sample points? Was I able to use the data to draw a map? If not, why not? | | |
| **5** Reaching conclusions | Was I able to reach firm conclusions from my data? If not, why not? Could I rely on any secondary data? | | |
| **6** Evaluating the enquiry | Did everyone use the data collection method in the same way so that I got reliable results? What might have happened if I had collected data at a different time of day or week? | | |

**Learning objectives**

■ How to think critically about each part of the fieldwork process.

# Evaluating data presentation

There are lots of different types of graph you can draw but choosing which graph to draw is not just about personal taste. There are some rules about which graphs you should use. So, when you are revising, you need to think about:

■ whether you chose the most suitable type of map or graph;

■ the strengths and weaknesses of the maps and graphs you have drawn;

■ if you were going to draw one of them again, how you might do it better.

A Lotus Diagram is a simple way to break down a big process (like **data** presentation of your fieldwork data) into smaller chunks. To draw a Lotus Diagram use the following steps.

**Step One** Draw a grid of nine squares in the centre of a large sheet of paper.

**Step Two** Write the process that you want to evaluate in the middle square.

**Step Three** Now break that process down into eight smaller chunks and write those down in each of the eight outer squares. You could do this step in discussion with friends in class. See Figure 5.

**Step Four** Colour each of these eight squares in a different colour.

**Step Five** Draw another eight grids, each with nine squares. Write down each technique or factor from your first grid in the central cell of each new grid. One example is shown in Figure 6.

**Step Six Evaluate** how well your fieldwork went in each of these areas. Again, you could do this as a discussion with friends in class. Remember to think about strengths and limitations of the techniques you used. Try to think of eight different statements. It isn't essential to fill every cell in the grid. An example is shown in Figure 7.

**Figure 5** Eight aspects of processing and presenting arranged around the central part of a Lotus diagram.

| Photographs | Transects and cross sections | Numeracy |
|---|---|---|
| Text | How was evidence processed and presented? | Maps and graphs |
| Perceptions | Making connections | Change over time |

## Activities

1 Choose one presentation technique that worked well. Justify why you chose it. Think about the suitability of this technique for the type of data.

2 Choose another technique that worked less well. Evaluate this technique making sure you describe its strengths and weaknesses.

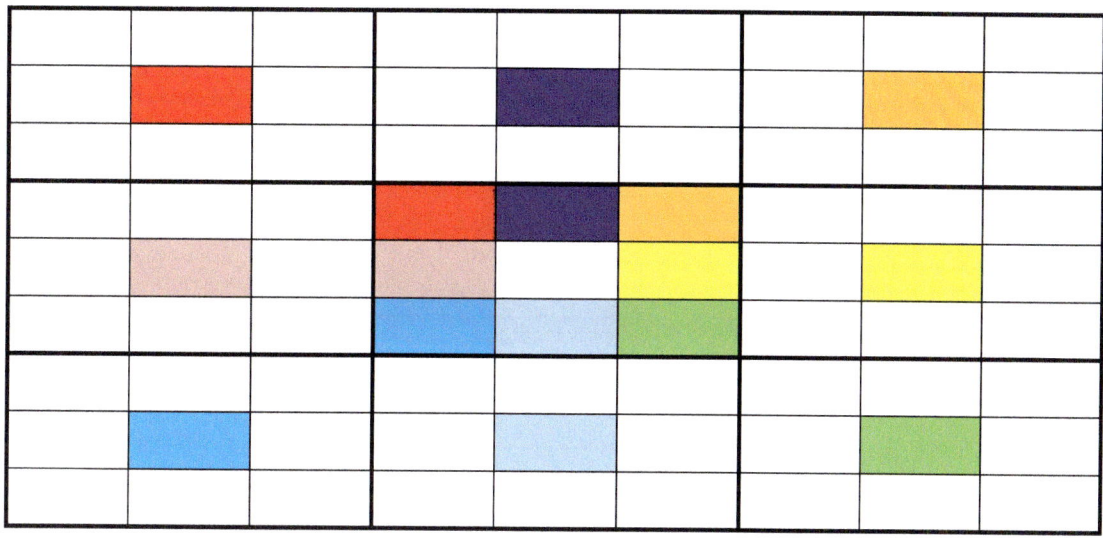

**Figure 6** The complete Lotus Diagram.

| If we had recorded where each respondent lives we could have mapped the results to see any **spatial patterns** | Lots of open questions in the **questionnaire** were time-consuming to process | I used word clouds to analyse responses to **open questions** which proved that some positive adjectives were used by many |
| --- | --- | --- |
| Processing the **pilot survey** gave unexpected results and helped create a better questionnaire | **Perceptions** | I used a **line graph** to process the **Likert survey** scores. With hindsight, this was unsuitable because the data is not continuous |
| It was difficult to process the positive and negative numbers in the **bipolar** scores | Bipolar scores were represented with **bar charts** because this is discrete data | **Mean** bipolar scores were calculated for different age groups |

**Figure 7** An example of a completed Lotus Diagram which evaluates one aspect of data processing and presentation.

# Fish bone diagrams

Fish bone diagrams like Figure 8 are a useful way to analyse the factors involved in any process. Drawing a fish bone diagram should help you to evaluate each factor that may have affected your **data** collection. The first thing to do is to list all of the factors that might have had any effect on how data was collected – these become the bare bones of your diagram. Then, you can think about the strengths and weaknesses of each of these factors. It may help you to identify some simple ways that your data collection could have been improved.

**Figure 8** Use a fish bone diagram to think about the strengths and weaknesses of each factor that contributed to the process of collecting the data.

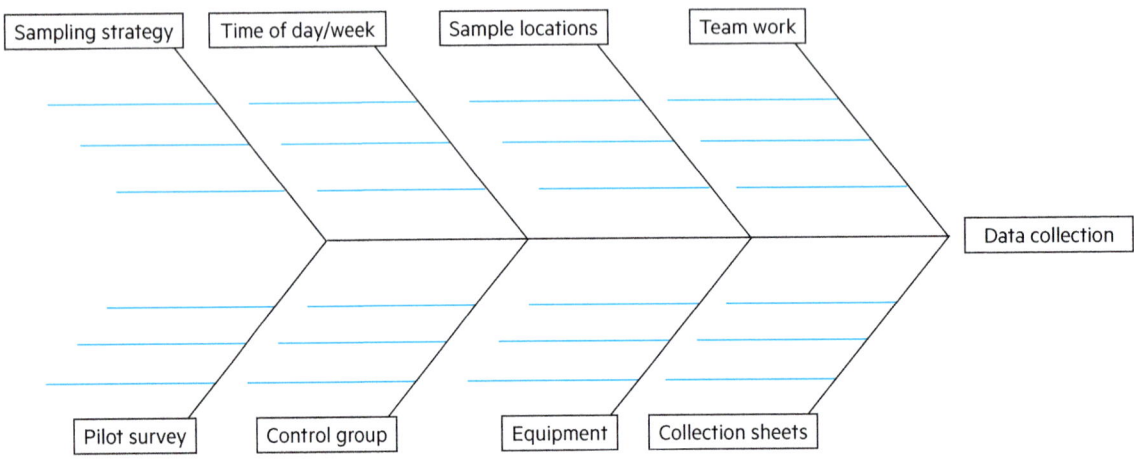

**Figure 9** Key words to use when you evaluate your fieldwork enquiries.

# The importance of evaluation

About two thirds of the marks in the assessment are for your ability to **evaluate** your fieldwork, or justify your decisions. Figure 9 shows some words that are useful when you are evaluating your fieldwork. Use these words whenever you discuss your fieldwork investigations.

## Activities

1  Discuss the sampling strategy you used to collect your primary data. Make sure you understand why it was chosen and make some notes that justify this choice.

2  Study Figure 8.
   a) Make a large copy of this fish bone diagram. Keep the main heading of 'Data collection' but think carefully about the factors – the headings at the end of each main bone. Most of these headings should be appropriate for your own fieldwork but if some aren't you should replace them.
   b) Working in pairs, discuss the strengths and weaknesses of each factor. Summarise your discussion by adding notes to each minor 'bone' of the diagram.

# Exam questions

Exam questions can be about any of the six stages of the enquiry process. There are two types of question about fieldwork in the examination.

- Some questions will be based on resources that you will see for the first time in the exam.
- Other questions will be about your own experience of fieldwork.

Questions about your own fieldwork might ask you to:

- **justify** a decision. For example, by asking why a fieldwork technique that you used was suitable or reliable;
- **evaluate** your fieldwork. For example, by asking you to identify the strengths and weaknesses of one way you used to represent data;
- **interpret or analyse**. For example, by asking you to suggest why the location used for your fieldwork was suitable;
- **make a judgement**. For example, by deciding which of two data collection techniques was most effective.

Avoid lengthy descriptions of your fieldwork. You need to answer the question so, read the command word and make sure that you use words that justify or evaluate what you did. It's a good idea to be direct when you are writing your answer. So, for example, if you are evaluating something then vague words like **helpful** or **useful** are OK but evaluative words like **advantage** and **disadvantage** are more direct. It's also a good idea to use qualifying words like **major** or **substantial** when you evaluate. This shows you have thought about the relative importance of the strengths and weaknesses. Figure 10 suggests some useful words to use in any evaluation. Another good idea is to use examples to make the evaluation specific to the actual fieldwork that you did. Figure 11 gives an example.

| Positive evaluation | Negative evaluation | Adding emphasis | Qualifying your evaluation |
|---|---|---|---|
| Advantage | Challenge | Chiefly | Significant / Insignificant |
| Benefit | Disadvantage | Especially | |
| Opportunity | Failure | Mainly | Major / Minor |
| Plus | Limitation | Mostly | Partial / Substantial |
| Strength | Minus | Particularly | |
| Success | Obstacle | | |
| | Weakness | | |

Figure 10 Words for an evaluation.

Figure 11 Refer specifically to your own fieldwork.

When I sampled housing in Shrewsbury I used random number tables to choose locations on an OS map. → One issue is that OS maps are sometimes out-of-date. → A major limitation of my sampling strategy was that the Darwin's Walk housing estate was so new it wasn't on the OS map so I didn't sample it.

This is a description which provides useful context. It's nice and short.

This is a weak evaluation and we don't learn anything about the actual fieldwork.

This evaluation is better because 'significant weakness' is more direct. It also uses a specific example from an actual piece of fieldwork to support the evaluation.

# Glossary

**Accuracy** How close a measurement is to the true value of the object being measured.

**Aim** What you hope to achieve or prove through fieldwork.

**Altitude** Height above sea level.

**Analysis** Process in which you make sense of the evidence.

**Anemometer** Fieldwork equipment used to measure wind speed.

**Annotation** Text added to a photo or artwork to analyse and highlight key features.

**Anomaly** Data that does not fit into the general pattern or trend.

**Assess** To use data and evidence to make a judgement.

**Average** The central point of a data set. The average can be expressed in three ways:

**Mean** Calculated by adding up all data in a set of data to find the total then dividing that total by the number of items of data.

**Median** The middle value when all data in a set of data is arranged in rank order.

**Mode** The most frequently occurring value in a set of data.

**Axis/y/x** The x-axis is the horizontal base line of a graph. The y-axis is the vertical line.

**Bar chart** A way to represent data. Values are shown using vertical columns or horizontal bars.

**Divided bar** A rectangle which has been divided into segments to represent data in the form of percentages.

**Base map** A simple, outline map of a place. Data recorded during fieldwork can be added to the base map.

**Beach profile** A cross section showing the landscape of a beach, both above the water and below it.

**Belt transect** (continuous & interrupted) A sampling strategy. Data is collected along a line (or transect) usually by using a quadrat.

**Bias** A tendency to be positive or negative about something or someone.

**Big data** Large sets of data and databases of secondary data such as the National Census.

**Bipolar survey** Data collection method in which people express an opinion by choosing from opposite pairs of statements.

**Break in slope** A place along a transect where the gradient of the slope suddenly changes.

**Causality** The relationship between something that happens and the thing that makes it happen.

**Census** A national database about the UK population that is collected every 10 years.

**Choropleth** Type of map where darker shading is used to represent higher data values.

**Clinometer** Fieldwork equipment used to measure gradient / slope angle.

**Closed questions** Questions with specific set answers so the person answering has to choose the answer they most agree with.

**Cluster(ed)** Group of similar things gathered or occurring closely together.

**Conclusion** A summary statement or judgement that summarises the findings of a fieldwork enquiry.

**Conflict matrix** A way to record potential conflicts, for example, between different uses of a place or space.

**Connection** The link that exists between one set of data and another.

**Control group** A group of data that acts as a benchmark. A researcher can compare the results from the sampled group to the control group to help identify the impact of different variables.

**Control readings** Measurements that are taken so that the effects of one variable can be isolated from an investigation.

**Correlation** The connection that links two sets of data (or variables) together.

**Negative correlation** Values in one set of data increase as values in the other set of data decrease.

**Positive correlation** Values in both sets of data increase at the same time.

**Counting** To record how many whole units there are.

**Cross section** Represents the shape or profile of a feature using measurements of the distances and depths from a horizontal line.

**Data** Facts and statistics collected together, for example, as evidence and for analysis.

**Big data** A very large and complex set of data.

**Bivariate data** Two sets of data that are connected in some way.

**Census data** Data about the whole population which is collected every 10 years.

**Continuous data** Values that can be measured and recorded to one or more decimal places.

**Discrete data** Values that can be counted and recorded as whole numbers.

**Primary data** Data that is collected first hand.

**Qualitative data** Evidence that is collected as words, opinions, or images.

**Quantitative data** Evidence that is collected as number values.

**Secondary data** Data that has already been published in another source.

**Dataset** A group of values (set of data) that is often presented in a table.

**Desire line map** Uses thin straight lines to show how places are linked together.

**Discharge** A measure of water flowing in a river. Discharge is calculated by multiplying cross sectional area of the channel by the river's velocity.

**Dispersion graph** A way to represent data. Data is plotted to show the range of values.

**Dot map** Uses individual dots to show the exact location of similar features on a map.

**Elaborate** To write an extended answer that links ideas and often uses connectives like 'so', 'therefore', 'because'.

**Enquiry process** The process by which a student collects and analyses evidence in order to make a decision or prove/disprove an aim.

**Enquiry question** A question that can be posed at the beginning of a fieldwork investigation to give the investigation an aim.

**EQI (Environmental Quality Index)** A technique which uses detailed criteria (statements) to assess the quality of our surroundings.

**Evaluate** To weigh up the strengths (or advantages) against the weaknesses (or disadvantages).

**Evaluation** A process in which you weigh up strengths and limitations. Evaluation is an important stage in your fieldwork enquiry in which you assess how well your fieldwork has gone. For example, you can evaluate the methods you used to collect or represent data.

**Evidence** Information that can be used to investigate an aim, test a hypothesis, or support an argument. Evidence may include quantitative or qualitative data.

**Extrapolation** A process by which a graph is used to find a data value that lies outside (or beyond) other data values.

**Field sketch** A sketch made during the fieldtrip which records evidence about the most important features of a fieldwork location.

**Fieldwork** Practical work undertaken in natural and urban environments to investigate geographical questions and hypotheses.

**Flow line map** A way to represent data. The amount of a flow (such as traffic or pedestrians) is shown using a proportional arrow.

**Flow meter** A piece of equipment to measure the velocity of the flow of a river.

**Footfall** A measurement of the number of pedestrians.

**Frequency** The rate at which something happens over a particular period of time.

**Geological map** A way to represent data. Rock types are shown on a map by using different colours.

**GIS** Geographic Information System. A system which captures, stores, analyses, and presents types of geographical data.

**Google street view** A feature in Google Maps and Google Earth that provides panoramic views from positions along many streets in the world.

**Grid reference** A reference which describes a location on a map via a series of vertical and horizontal grid lines identified by numbers and letters.

**Histogram** A type of graph that uses bars to represent continuous data.

**Hypothesis** A statement which can be proven to be correct or incorrect based on the evidence collected in your fieldwork.

**Infiltration (rates)** The speed at which rain water soaks into the soil.

**Interpolate/interpolation** A process by which a graph is used to find a data value that lies between other data values.

**Interquartile range** The middle 50% of a data set. The range from Q1-Q3.

**Interview** A way of collecting qualitative data. Interviews involve asking open questions. An interview is usually less structured and may be more detailed than a questionnaire.

**Investigation** Using enquiry skills to test a hypothesis or answer a question by collecting and analysing data and evidence.

**Irregular** Not regular. Does not fit a pattern or set of rules.

**Isoline map** A way to represent data. Data points with the same value are joined by a line.

**Judgement** A decision that you must support by use of evidence.

**Key** The legend of a map or graph that describes the meaning of colours or symbols that have been used.

**Kite diagram** A way to represent data.

**Line graph** A way to represent data. Values are shown by drawing a line that joins the data together.

**Linear** Arranged along a straight or nearly straight line.

**Likert survey** A data collection method in which people express an opinion by stating how strongly they agree or disagree with statements.

**Logarithmic** A scale on a graph where the numbers increase more and more rapidly.

**Mean** A measure of central tendency (or average) in a dataset.

**Measuring/measurement** The action of calculating the size, length, distance, area, quantity, depth, range of something.

**Median** A measure of central tendency (or average) in a dataset.

**Mode/modal class** A measure of central tendency (or average) in a dataset.

**Open questions** These allow the person being interviewed to say whatever they want in response.

**Opportunistic** A sampling strategy. Data is sampled where it is convenient to do so.

**Ordnance Survey (OS) maps** Online and paper maps that show the landscape features of the UK.

**Origin** The place on a graph where the two axes meet.

**Pattern** A regular form or sequence that can be seen.

**Perception** What someone thinks of a place, idea, situation.

**Pictogram** A way to represent data. A simple picture is used to show the value of data.

**Pie chart** A type of graph where a circle is divided into segments that each represent a proportion of the whole.

**Pilot survey** A stage of data collection that allows sampling methods to be tried out and evaluated before a full sample is taken.

**Profile** An outline that represents a feature as seen from one side.

**Beach profile** A diagram that represents the gradients of a beach.

**Long Profile** A diagram that describes how a feature such as a river varies along its length.

**Proportional symbol map** Uses symbols (usually squares or circles) of different sizes to represent data.

**Quadrat** A piece of fieldwork equipment. A square frame that is used to sample data (often vegetation).

**Questionnaire** A way of collecting qualitative data that involves asking a number of respondents a common set of questions.

**Random number table** A table of digits (0 to 9) arranged in an unpredictable sequence in rows and columns and used to create a random sample.

**Range** Measure of dispersion – the difference between the highest and lowest value in a dataset.

**Rank order** A process by which numbers are put into sequence by their value from largest to smallest.

**Reliability** The degree to which the evidence collected during an investigation can be considered consistent and dependable.

**Re-photography** The process of taking a photograph now (primary data) and comparing it to a photograph or other image taken in the past (secondary data).

**Representative** Sampled data that is a true reflection of the whole.

**River discharge** Amount of water flowing per second down a river channel.

**Sample** A small part or quantity that can show what the whole of something is like, for example, a group of people drawn from the population of an area used to show what that population is like.

**Sampling** The process of selecting a sample.

**Random** Data is sampled at irregular intervals.

**Regular** Data is sampled at equal intervals.

**Stratified** Data is sampled in proportion to the size of the original population.

**Systematic** Data is sampled at regular intervals.

**Transect** Data is sampled along a line.

**Scale line** A visual way of showing distance on a map by drawing a short line to a known length.

**Scatter graph** A way to represent bivariate data. Data points for two variables are shown on the same graph.

**Sketch** A simple drawing that represents the main features of a landscape.

**Sketch map** A simple plan that represents the main features of a landscape.

**SMART** A statement, aim, or objective which is Specific, Measurable, Achievable, Realistic, and Timely.

**Spatial pattern** How features are distributed across a map.

**Species** A group of living things with similar characteristics.

**Surveying** The process of observing or measuring features in the environment and then mapping or recording them in some way.

**Tally mark** A quick way to keep track of numbers. A vertical line equals one. Tally marks are made in groups of five. One vertical line for each of the first four numbers and a diagonal line for the fifth.

**Text analysis** Looking for patterns or trends in words.

**Transect** An imaginary line through the environment. Data and observations can be made along this line.

**Trend** A change shown on a graph.

**Unreliable** Data that has been collected using a technique that is inconsistent.

**Variable** A feature, element, or factor which changes.

**Dependant variable** A variable whose value depends on the value of another factor (such as distance, time, or height).

**Independent variable** A factor that causes other data to vary.

**Vegetation** Plants, shrubs, trees.

**Velocity** The speed at which something flows.

**Ward** A small rural area or a neighbourhood of a town or city. Wards vary greatly in area and in population but on average each ward in England and Wales contains 6,600 people.

**Weighting** A way of giving greater importance to some of the data that is collected.

**Word cloud** A visual way to present the most commonly used words in a piece of text.

**x-axis** The horizontal base line of a graph.

**y-axis** The vertical line of a graph.

# Index

Index